T0178159

Product Development

Christopher A. Mattson • Carl D. Sorensen

Product Development

Principles and Tools for Creating Desirable
and Transferable Designs

Christopher A. Mattson
Department of Mechanical Engineering
Brigham Young University
Provo, UT, USA

Carl D. Sorensen
Department of Mechanical Engineering
Brigham Young University
Provo, UT, USA

ISBN 978-3-030-14901-7 ISBN 978-3-030-14899-7 (eBook)
https://doi.org/10.1007/978-3-030-14899-7

This Springer imprint is published by the registered company Springer Nature Switzerland AG.
The registered company address is: Gewerbestrasse 11, 6330 Cham, Switzerland

Preface

In preparing to write this book, we reviewed more than two hundred design-related textbooks. With all the existing books, why did we decide to write *this* book?

In more than twenty years of teaching engineering design classes, covering more than 700 industrially sponsored design projects with more than 4,500 students, we've come to believe that a book could be written to better match the needs of both our student teams and practicing design professionals.

In reviewing the design textbooks that are currently in print, we've identified three broad classes of content: (1) the design process; (2) design tools, methods, and techniques; and (3) design-related topics such as patent law and ergonomics. The differences in the books are largely in the depth and breadth of each of these categories, although there are differences in presentation as well (e.g., the number and detail of the examples or case studies that are used to teach the main topics). The books are largely consistent in focusing their discussions of the design process on the activities of the designers during different phases of the process.

In our capstone class, each team has a different project. Although all projects require engineering design, they vary significantly in scope, expected results, and involved engineering disciplines. For example, one year we had teams designing vehicles, improving a potting process for X-ray tubes, improving the packaging and distribution process for a nutritional supplement company, designing the next-generation extension ladder, and developing a plan to prevent the release of visible water vapor from a cooling tower in a copper smelter. Each of these projects requires a design process that is customized to the particular product being designed. Yet the existing textbooks give little or no guidance as to how a process might be customized. As experienced designers, we know how to customize the process. But our inexperienced students have yet to learn how to do this. Therefore, when we talk about the generic design process as a set of design activities that will apply to a typical product, virtually every team can claim that traditional product development processes don't apply to their specific product.

Furthermore, the idea of a design process being fundamentally described by the activities currently being pursued by the design team doesn't square with our experience in industry. When discussing product development projects, team members are not very likely to classify the product by their actions. Rather, they tend to describe the characteristics of the product. We believe that describing the design process by the evolution of the product is clearer, more easily understood, and more consistent with the way experienced designers think about their work.

A fundamental shift in this book, then, is away from the focus on prescribing the work of the designer toward a focus on the evolution of the product. This shift in focus is both challenging and liberating to the students (and by extension, to design professionals). The challenge comes because we no longer prescribe actions; instead, we prescribe outcomes. The student must determine the actions that will best lead to the desired outcome. For an inexperienced designer, this can be hard work, fraught with uncertainty. And yet, for all its difficulty, this work is innately rewarding, as the student successfully wrestles with issues of real importance.

The liberation in the focus on product evolution comes from the same source as the challenge: we no longer prescribe actions. Students feel no pressure to follow a particular action because it is part of the recommended process. Instead, they need only focus on actions that are appropriate for their particular design project. For the industry professional, this book can serve as a "field guide" for product development. It provides concise guidance about both the activities and the outcomes of product development. The book also presents effective development tools and techniques for the real world.

Having tested these principles in our senior capstone design class, we have found that they work. Students learn better, work better, and accomplish more when they focus on the evolution of their product rather than on an accounting of their activities. They are much more likely to consider instructor guidance as suggestions to help them reach their desired product evolutionary state. Furthermore, they appear to better understand and are certainly better able to articulate the reasons why they are involved in a particular activity.

For the industry professional, this book can serve as a "field guide" for product development. It provides concise guidance about both the activities and the outcomes of product development. The book also presents effective development tools and techniques for the real world.

Part I of this book covers the fundamentals of the product development (or design) process as we have come to understand it. We introduce four principles that underlie product development, including the importance of demonstrated desirability and transferability for product designs. We explain how design skills are applied in an incremental and iterative fashion to cause

the evolution of the design. We introduce design activities, outcomes, and the activity maps that link them together. We describe the evolution of design information as it applies to all product development projects. We show the importance of coordinating the work of all team members and demonstrate how an activity map forms the basis for customizing the development process to the needs of the team and the project. Finally, we discuss some personal characteristics of designers that we have seen influence success.

We also include a case study of product development. Through showing how a team successfully dealt with the unique challenges of their assigned product development opportunities, we hope to give some concrete guidance and inspiration to those who are struggling with their own project.

Part II of the book is a Product Development Reference, a brief reference to product development tools, methods, and techniques. The Development Reference is certainly not exhaustive, either in coverage or in depth. However, we have attempted to include a reference entry for each of the tools, methods, or techniques that is mentioned elsewhere in the text.

Only the first part of the book, the product development process, is intended to be read consecutively. The remaining parts can be browsed through or cross-referenced from Part I. We hope that the book will be inviting to read, and friendly to review, and that it will provide insight and inspiration not only during a classroom design experience, but throughout a long and successful career in engineering design.

We thank WHOlives.org for allowing us to share the story of the Village Drill (human-powered water well drill) development. They've helped improve the world not only by bringing clean water to more than a million people in Africa with the Village Drill but also by providing an example of the application of the principles in this book to help improve product development.

We gratefully acknowledge Nichole Cross for providing a style guide to achieve consistency in the figures and for her capable creation of the illustrations in this book. Without her help, the book would be far less readable.

Provo, UT, USA   Christopher A. Mattson
                                          Carl D. Sorensen
August 2019

Contents

List of Figures

List of Tables

Part I

Product Development Process

## CHAPTER 1

Getting Started

1.1 On the Shelves

I'll never forget the first time I saw something I designed on the shelves of a store[1]. It was a simple docking station for a handheld computer (Fig. 1.1). When I saw it, I picked it up and marveled at every piece of plastic, every screw, and every layer of plating on each of the precisely formed electrical contacts. It was one of those rare and embarrassing moments where I'm sure I heard dramatic music from a movie soundtrack playing in the background. To say I was proud about what was on that shelf would be a huge understatement.

Certainly no one else paused to admire the product's finer details, nor did anyone buy the product because of the plastic wall thickness, or the screw diameters, or because of the plating on the contacts. But somehow — and not by chance — all of those things (and many, many more) worked together to form a product that was desirable enough for people to want to buy, and so desirable that we ended up selling over 12 million of them.

It was not the first time I had seen the product — that very one in the store could have been one of hundreds I saw come off

the production line weeks earlier in China. But when I saw it in that store — away from my desk, away from the computer models, and away from the to-do lists — I began to see the big picture. My understanding of product development began to mushroom, as did my love for it. I could see then, with better clarity than before, the importance of the seemingly endless number of decisions that went into designing the product. I could see the many people who were involved in its development; what they worked on versus what I worked on and how we often helped each other out in various ways. I could see that the product had evolved little-by-little — and not in a haphazard way — from a simple idea to a manufactured product that people could now buy at stores.

* * * * *

It would be misleading to imply that I had always had a good relationship with the docking station. The truth is that it caused a significant amount of stress for me and others on the development team.

I had been working for the company as an engineering intern for about two years when I was assigned the project. Until that point, I had spent most of my time troubleshooting problems and developing reliable test methods for other products. To my boss, perhaps this new assignment was

---

[1] This is Chris's personal experience.

© Springer Nature Switzerland AG 2020
C. A. Mattson, C. D. Sorensen, *Product Development*,
https://doi.org/10.1007/978-3-030-14899-7_1

Figure   1.1:
Docking station
for   handheld
computer.

no different. The docking station was in the very early stages of product development, and despite the best of efforts, the docking station's most critical component — the custom electrical connector system — was not looking promising. My job was to fix the situation and get a working design ready for production.

Things were urgent, as they often are, and within hours my boss and I were on a plane headed to work in the Taiwanese office, where some of the other team members were based.

I'll be honest; as a good engineer, I could analyze many things. But I had no clue how to create something from scratch. I knew basically nothing about product development — and it scared me.

In an attempt to remedy this, I studied a textbook on product development for the entire flight to Taiwan. Although that book did not solve my problem, I was able to extract what I absolutely needed to get started on developing this product. The things I walked away with from that study not only gave me the much-needed confidence that ultimately led to a

successful product, they also led to the creation of this book, which is designed to help anyone who is in the same or similar situation — charged with the task to develop a product, but not quite sure how to do it, or how to do it well.

What ultimately sprang from that study and from the development of many products (including the docking station) was simple, but very powerful — it was an understanding of what happens to the design during product development and what the team does to cause the development to happen.

1.2   What This Book Offers

This book offers precisely what I needed — and in the format I needed it — while on that plane to Taiwan. It describes what happens to a design as it evolves through distinct stages of development from a simple idea to a product on the shelves. Importantly, it will help you understand what stage of development your product is currently in, and what you and your team need to do to advance it to the next stage. In other words, it describes how to design a product from scratch — and not just any product, but a product the market loves.

While it would have been valuable for me to have this book while designing the docking station, it is equally valuable for the design of *any* engineered product. How do I know? Because Carl and I specifically created this book to be universal *after* we had observed, interviewed, taught, coached, and/or participated on hundreds of product development teams. And we verified the universality of the book by testing it on dozens of development teams in many settings.

The Layout

We've created the book in two parts — Part I: Product Development Process, and Part II: Product Development Reference.

Part I discusses the process of product development. Specifically it focuses on the stages of product development and how to manage individual and team efforts during those stages so that it results in a product the market really wants.

Part II is an alphabetized reference that introduces standalone actions, methods, and tools that have the potential to help the product development team advance the design. We present them in a reference-book style because they are often used throughout product development and are not structurally tied to any one chapter in Part I. Throughout the book, Part II is referred to as the *Product Development Reference* or the *Development Reference*.

Part I also includes a summary of key product development information (page 309), a detailed case study (page 327), and a glossary of product development terms (page 341). Generally, it's worth giving Part I and the appendices a full read, while following citations to the Product Development Reference when you're unfamiliar with a particular action, method, or tool, or when you simply want to understand our perspective on it.

Because the Development Reference provides the necessary details to carry out product development, we anticipate that it will be turned to as needed and that it will

provide you a place to be introduced to a particular design action, method, or tool that will be of immediate help.

1.3 Four Principles of Product Development

The concepts, methods, and discussion presented in this book assume that you understand the Basic Design Process (as described in the Development Reference: Basic Design Process (11.1)). Furthermore, all of the content in this book is based on four principles that define the context for product development. While simple and logical, these principles and their deeper meaning are rarely understood by those just starting with product development. A solid understanding of them can help you avoid wasting product development time, and can help you avoid frustration.

Principle 1: The job of the product development team is to create a design that is desirable and transferable

The job of the product development team is to create a design (a clear and complete definition of the product), while the job of the production system is to use that design to manufacture the product in quantities consistent with the market demand.

Recognizing this subtle distinction between jobs will help you focus on what you're expected to create as a product development team, and to know how your work will be judged. It will help put into perspective the critical — yet transitionary — role that prototypes play in advancing the design from one state to the next, and help you avoid the common trap of believing that your only job is to produce a prototype.

To be successful, the product development team will need to create a design that is both *desirable* and *transferable*. This principle is represented by Figure 1.2.

The goal of desirability assumes that for a product to succeed in the marketplace, it must be desirable, meaning that the target market wants the product enough to purchase it. A desirable product results

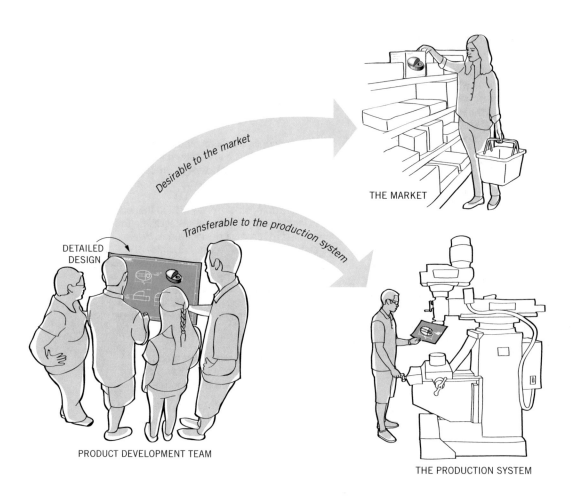

Figure 1.2: Principle 1: the job of the product development team is to create a design that is both desirable and transferable. Notice that the job of the production system is to produce the product according to that design.

when a high-quality production system manufactures the product according to a desirable design.

The goal of transferability is that the design be transferable, meaning that anyone producing the product in accordance with the design will create a product that matches the intent of the development team in all important ways. To do this, the production system needs to receive a design that is both *clear* and *complete*. A design is clear if it is unambiguous, meaning that no judgment is required to interpret what is meant by the design. A design is complete if all important aspects of the product are included in the design. If the design is clear and complete, it is impossible for a production system to create a product that is simultaneously consistent with the design and inconsistent with the intent of the development team.

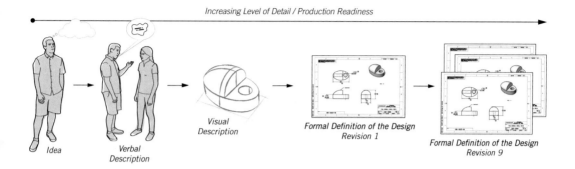

Figure 1.3: Principle 2: The design must evolve, gradually becoming better, more mature, and until it contains all the necessary information for the production system to manufacture the product and test its quality. Representative stages of evolution shown here include the idea, a verbal description, a sketch, and multiple revisions of the formal technical definition. There are many intermediate stages that could be shown.

The desirability and transferability of the design should be regularly evaluated during the product development process to ensure the process is on track.

Throughout this book we focus on the importance of desirability and transferability because when designs are both desirable and transferable, the resulting products succeed in the market.

Principle 2: The design must evolve, gradually becoming better, more mature, and until it contains all the necessary information for the production system to manufacture the product and test its quality

Designs evolve little-by-little from an embryonic state to a completely defined state during the product development process. The principle of evolution is illustrated in Figure 1.3. In fact, the successful evolution of a design is the sole purpose of product development. Importantly, the evolution of the design has been successful when it results in a desirable and transferable design. As such the design evolves in two important ways; it becomes better (more desirable) and it becomes more detailed (a clearer, more complete, transferable definition of the product).

Understanding the principle that designs evolve can help you avoid frustration when a desirable design does not appear all at once. It will help you plan for the desirable and transferable design to emerge gradually as a result of hard work and iteration. Importantly, this principle helps you recognize that in the end, the design will need to have evolved to the point that it includes all the necessary information for the production system to manufacture the product and test its quality. Without this, the product development effort is incomplete and requires additional work.

Principle 3: The product development team causes design evolution through design activities that result in artifacts

The gradual and consistent evolution that occurs during the product development process only happens as a result of actions taken by the product development team. This is represented in Figure 1.4. The evolution is often a result of iteration — meaning as a result of repeating some product development activities as a means to get closer and closer to a desirable design.

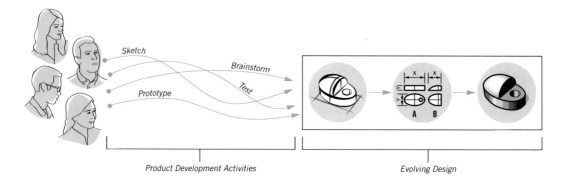

Figure 1.4: Principle 3: the product development team causes design evolution through design activities that result in artifacts.

And not just any actions result in desired evolution. In order to cause the design to evolve, the transferable representation of the design must change. We refer to elements of this transferable representation as *product development artifacts*. Product development artifacts move the design out of a designer's head and into a tangible form that is transferable. Hence, evolution requires activities that result in artifacts.

Understanding this principle is both empowering and frightening, as it indicates that the product development team is free to choose what it should do to cause evolution, but it also indicates that those choices will either make or break the product.

The bulk of this book is about the stages through which a design passes while it is evolving, and about the activities the product development team must do in order to move the design through the stages of development.

Principle 4: Optimal evolution requires customization and coordination of activities

The product development process is most efficient and effective when the product development activities are coordinated and customized to the unique conditions of the product, the client, and the team.

As we seek to evolve the design, we will carry out many product development activities (Principle 3). Those activities must be chosen and coordinated for the purpose of converging (Principle 2) on a design the market finds desirable, and the production system finds transferable (Principle 1). Figure 1.5 is an illustration of Principle 4, showing a generic set of design activities coordinated amongst the people doing them. These activities are arranged purposefully, leading to a tested design. Without purposeful coordination, the actions of multiple people can easily cause waste, duplication, and delay as shown in Figure 1.6.

Because every product is different, and every client has their own needs, and because every product development team has its own strengths and weaknesses, the product development activities that *should* be chosen, and the way they *should* be coordinated (or sequenced) will be different in each case. For this reason, product development will be most effective and most efficient when it is customized to the unique characteristics of the product, client, and team.

Customizing product development in this way is not as daunting as it sounds. A major goal of this book is to equip you with the information and tools necessary to customize product development to any setting you find yourself in. As we've

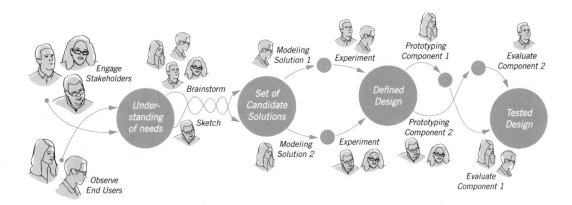

Figure 1.5: Principle 4: optimal evolution requires customization and coordination of activities. Notice that this purposeful coordination of activities among the people doing them leads to a valuable outcome.

observed designers acquire this skill, we've seen less floundering and less frustration. We have also seen industry recognize this as one of the designer's most valuable skills and hire them because of it. There is a marked difference between a designer who can only follow a prescribed process, and one who can prescribe the process best suited to the unique setting of his or her development team.

* * * * *

Everything in this book builds on these four principles. They are the starting point for everything we'll discuss. Keep an eye out for them while you go through the book. Sometimes they'll be stated explicitly, and other times they'll be quietly at the root of the subject.

Looking back now — at that moment in the store when I saw the docking station for sale — it's clear to me that the four principles discussed above are a fundamental part of that story.

I stood there — in that store — holding a product that people were buying; not one I manufactured, but one that our product development team fully defined so that it could be manufactured by a production system! That's Principle 1.

How about evolution? That's Principle 2. The excitement I felt seeing the docking station on the shelves was in large part because I had nurtured the design through various stages of development as it evolved into the final product. I remember planning to solve the entire problem in just a few days while in Taiwan! I hadn't yet learned or accepted that great designs only emerge after a lot of iteration.

In fact, we did hundreds of product development activities aimed at creating and verifying desirability (such as systematically dropping dozens of docking stations from 6 feet onto concrete floors to identify weak points). That's Principle 3. We also did many activities centered on transferability. Not only did our product development team release (transfer) the design to our own production system, we also released it to our client's second-source; this was an important and welcomed part of our responsibility to protect the client against the risks associated with single sourcing.

As simple as the docking station seems, there was a sizable team that developed it. That team involved people of various disciplines, and companies higher up and lower down in the supply chain. I learned a lot about project coordination, not because

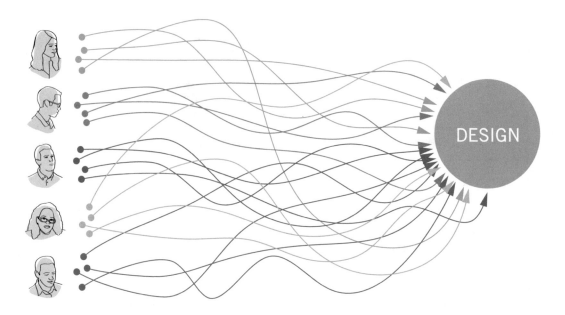

Figure 1.6: For real products, there are so many individuals and activities involved in the product development effort where coordination is required to avoid waste, duplication, and delay.

I did it, but because I saw great people organize the team and through that organization accomplish very difficult things. Those accomplishments are proof of Principle 4.

1.4   Moving Forward

By this point in the book we hope you can see that we intend to help you create a design that can be used by a production system to manufacture a product in quantities ranging from one to millions. Our focus, however, will not be on just any design — but instead on a desirable and transferable design. Creating such a design will not happen overnight, but it will happen little-by-little as you and your team carefully choose, coordinate, and customize your actions. Causing this evolution to happen is not easy — it wasn't for me and my team as we worked on the docking station, and it won't be for you and your team — but it is incredibly rewarding and it is certainly

made easier when you use the principles taught in this book to cut out the waste and avoid the pitfalls that are easy to walk into while developing a product.

1.5   Exercises

Test Your Knowledge

T1-1 Describe the two main parts of this book.

T1-2 List the four principles of product development.

T1-3 List the two main attributes of a design that should be regularly evaluated during the product development process.

Apply Your Understanding

A1-1 For a specific product development project with which you are acquainted, describe how each of the four principles of product development apply or do not apply to that project.

A1-2 Describe what might happen during the product development process if a development team believes that their job is to produce a product — as opposed to creating a design.

A1-3 Write some personal goals for a project you are working on now that might help you stay focused on your job to create a desirable and transferable design.

A1-4 Consider the deeper meaning of Principle 1. Can someone from the production system be on the product development team? Why would this be or not be a good idea?

A1-5 How can desirability and transferability be specifically defined for your development project?

A1-6 In what ways do you believe iteration can help you during the product development process?

A1-7 Where can you find out more about the stages through which a design will pass as it evolves?

A1-8 What does it mean to coordinate and customize product development activities? How is coordination different in an individual setting versus a team setting?

A1-9 What do you most want to remember from this chapter?

Product Development Fundamentals

2.1 Designing the Process

One of the first things to know about product development is that there are countless ways to do it, and what's even more interesting is that there are multiple right ways to do it.

To understand this better, consider how many products have been developed in the past 150 years (this number is enormous). Now imagine the designers of all those products, and the things they did to develop their product. Do you suppose they all did the same things, or followed the same process? Or even that the successful products followed one process, and the failed products followed another?

While it would be romantic for this to be true (not to mention extremely convenient), it unfortunately is not. There is no universal process for product development, no universal set of steps that apply to all settings, and there is no universal recipe of ingredients and quantities that can simply be followed to get a design that is both desirable and transferable.

But don't let this discourage you. There is actually something quite beautiful about it; something that gives our profession a deeper meaning. It is that we, as designers, have two great opportunities in front of us; the obvious one, which is to create a design

that is desirable and transferable, and the non-obvious one, which is the opportunity to choose the product development activities that work best for us, for our product, and for our client.

This means we'll not only choose the shape and material for a product (among other things), but we'll also choose the activities we will do to efficiently and effectively identify the *best* shape and *best* material for the product. These two opportunities have always been and will always be a part of what it means to be a designer and to develop a product.

In a way, this can be compared to a journey from one point on a terrain to another, by foot (Figure 2.1). Not only does the person traversing the terrain take the physical steps to get from the start to the finish, but she also plots the course, sets the speed, plans the places to stop and rest along the way, and makes judgments about when to check the map and reorient. Just like the designer, she also has two great opportunities: to *plan* the journey and to *take* the journey.

* * * * *

Our goal with this chapter is to help you understand fundamental aspects of product development that are common to successful development, and to describe

© Springer Nature Switzerland AG 2020
C. A. Mattson, C. D. Sorensen, *Product Development*,
https://doi.org/10.1007/978-3-030-14899-7_2

Figure 2.1:
Steps, paths,
and milestones
to the hiker are
like iteration,
sequences of
activities, and
stages of prod-
uct development
to the designer.

them in a way that will allow you to see them as building blocks for constructing the product development process you'll follow as you develop your product.

By the time you finish this chapter you should be familiar with three things. (1) the STEP cycle, (2) the activity map (sequence of design outcomes and activities), and (3) the stages of product development. These three things, in this order, represent an increasingly wider view of what happens during product development. When you understand them and how they work together in harmony, you'll be in a position to make the most of your time, and avoid wandering aimlessly.

To put these three things into perspective, let's again consider the hiker:

Footsteps — During her journey, individual footsteps will take her from the starting point to the ending point. There will be thousands of them and once she is comfortable with the gear, these stepss will become mostly automatic and

unconscious. As we'll describe in this chapter, to the designer these repetitive footsteps are similar to the STEP cycle that will be used to evolve the design. Like footsteps, the STEP cycle is the main engine of progress. The cycle is introduced in Section 2.2.

Paths — The hiker won't walk in a random unplanned way. She'll choose a general path from one point to another, then walk her way closer and closer to the goal. For the designer, the path is the sequence of design activities that will be executed as a way to evolve the design. Just as the hiker's path will be accomplished step-by-step, the designer will rely heavily on the STEP cycle while carrying out a sequence of design activities. This is described in Sections 2.3 and 2.4.

Milestones — The journey will be long, so she'll need to rest, reorient, and re-plan throughout the trek. She chooses milestones along the way to do this; scale that peak on the left, camp in that valley

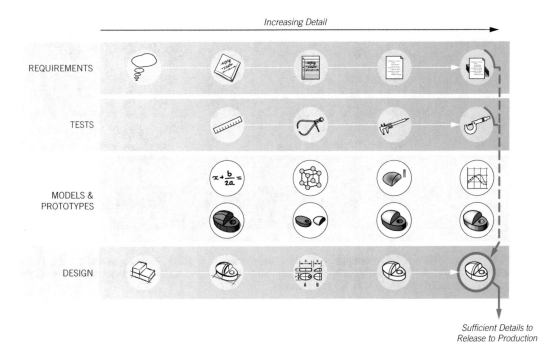

Figure 2.2: Design evolution during product development. The design evolves from an idea to a desirable, transferable design sufficient for use by the production system to manufacture the product. Along with the design, requirements and tests also evolve. Models and prototypes will be used with the tests to determine how well the design meets the requirements.

straight ahead, climb the peak on right, and so on. In this way, her difficult journey is divided into smaller, more manageable pieces with specific subgoals. To the designer, passing milestones is like completing stages of product development. We elaborate on these stages and how to benefit from them in Sections 2.5, 2.6, and 2.7.

2.2 Design Evolution and the STEP Cycle

Before a product can be manufactured and show up on the shelves of a store, its design must evolve (Figure 2.2). To be clear, the *design* is the deliberate definition of the product. In its initial form as an idea, the hope for a product exists in the mind of the designer, but this hope is not yet transferable to the production system that will make it. The design must evolve into a clear and complete (fully detailed) definition of a desirable product if the product is to be successful.

Along with the design, requirements and tests will also evolve during product development. Requirements clarify what is needed in order for the design to be desirable. Tests clarify how the performance of the design will be predicted and measured so its desirability can be evaluated by the development team. Without clear requirement and test information, the desirability of the product cannot be evaluated until it is actually placed on the market.

Prototypes and models play a central role in product development because they

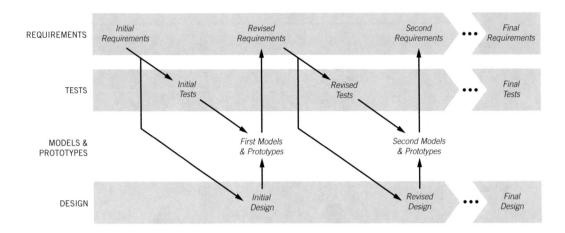

Figure 2.3: The co-evolution of the requirements, the tests, and the design as aided by prototypes and models. Proto-types and models are created as a snapshot consistent with the current design and tested to measure and predict the performance of the design to compare with the requirements.

represent the current design and are used to test how well it meets the requirements.

Figure 2.2 shows the evolution of the requirements, tests, and design. As depicted, it appears as if the evolution is consistent, easy, and smooth, much like driving on an interstate highway. In reality, design evolution is more like driving on surface streets — intermittent, difficult, and messy.

Clearly, we would all like to create the perfect design right from the start, but experience has shown us that this is impossible. Instead, we create a design that is the best we can do, then identify the weaknesses in our work, and refine the design. It's important to recognize that we'll do this many times before the design is ready to be used to manufacture the product. In this way, design evolution is both incremental and iterative.

It is incremental because design evolution occurs in discrete increments — not all at once. These increments are often relatively small changes to the design that can seem almost insignificant. But the cumulative effect of these increments is significant

evolution. Design evolution is iterative because basic design actions will often be repeated again and again to cause the next discrete evolutionary increment to be achieved.

Computer scientist Frederick Brooks (2010) describes the incremental and iterative nature of design well with a personal story:

> As a student I spent one summer working at a large missile company.... After a couple of weeks, I had a working [prototype of a data management system]. I proudly presented [it] to my client. "That's fine—it is what I asked for—but could you change it so that...?" Each morning for the next few weeks, I presented my client with [an improved prototype], revised yet again to accommodate the previous day's request. Each morning he studied the [prototype] and asked for yet another revision.... For a while, this frustrated me sorely: "Why can't he make up his mind as to what he wants? Why can't he tell me all at once, instead of one bit a day?" Then, slowly, I came to realize that the most useful service I was performing for my client was helping him *decide* what he really wanted.

Was Brooks part of an iterative process, or was the previous day's work simply undone

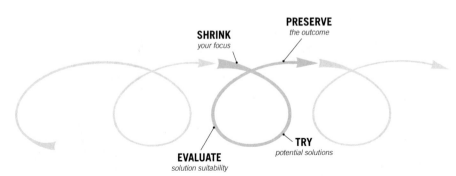

Figure 2.4: The STEP cycle: Shrink, Try, Evaluate, Preserve.

each day by an indecisive client? Unless we pay close attention to what's happening in this story, it's easy to believe that the incremental process was ineffective and inefficient, when in reality it was a healthy part of the evolutionary process. What Brooks experienced — the co-evolution of the requirements and the design — is a common and welcomed part of the product development process.

The client in Brooks's story provided insight that allowed Brooks to *add* to his understanding of the requirements, thus allowing him to improve the design day-by-day. The frequency of Brooks's interaction with his client was a large contributor to the success of this interaction.

The way in which the requirements, the tests, and the design evolve together during product development is illustrated in Figure 2.3. As shown, the initial requirements guide both the creation of the initial design and the establishment of the initial tests. The first prototypes and models are then made according to the initial design and are tested according to the initial tests.

The results of these tests will likely lead to revised requirements and tests, and will almost always lead to a revised design. The revised design is then used when creating a second set of prototypes and models, which upon testing is likely to kickstart another design iteration. This iterative process continues until the design has

evolved to a point of sufficient desirability and transferability.

Understanding this co-evolution makes it easier to avoid the frustration initially felt by Brooks, and helps the team see the value — even the essential role — of iteration during product development.

The STEP Cycle

Incremental and iterative design evolution is facilitated through an informal iterative pattern of action we call the *STEP cycle*. This basic four-part pattern, shown in Figure 2.4, underlies most of what is done during product development. Like the hiker's footsteps, this cycle is the main engine of progress. Its four parts are:

1. Shrink

   Shrink your focus to a manageable piece of the large problem. We do this for the simple purpose of making the large problem solvable. Just like the hiker's footsteps, our large journey is divided into multiple achievable pieces. The Decomposition entry in the Development Reference: Decomposition (11.14) can help you develop your ability to do this effectively.

2. Try

   In the second part of the cycle, we consider possible solutions and tentatively choose one that we hope solves the smaller more manageable piece. This may require various design skills (described more fully in Chapter 3). During this part of the cycle, we think about *how* the chosen solution does or would work.

3. Evaluate

   In this part, we evaluate *how well* the chosen solution solves the manageable piece of the larger problem. Based on the evaluation, we decide whether to Try additional solutions (in hopes of finding something better), or to Preserve the chosen solution by adding it to the design. In rare instances, a third alternative may be chosen: to abandon this STEP and move in a different direction entirely.

4. Preserve

   When a tentative solution has been evaluated and found worth keeping, the work is preserved by formally making the solution part of the design. To preserve the work, the design must be changed to reflect the outcome.

For most designers, this pattern of thinking (or some representation of it) permeates everything they do. For them, the pattern has been practiced enough, with enough success, that it becomes a natural and often unconscious way of operating. This is much like the hiker, whose individual footsteps are natural and unconscious, and in succession take her to her intended destination.

2.3   The Sequence of Design Activities and Outcomes

By themselves, the hiker's footsteps are not enough. She must first point herself in the right direction and chart a path to follow before her steps will lead to progress.

It's no different for the designer whose iterative (step-wise) efforts will have no meaning without a charted path to follow. To the designer, the path ahead is defined by the design activities that will be carried out to cause the design to evolve.

Compared to the relatively small-scale evolution caused by the STEP cycle, the evolution that occurs when executing a sequence of design activities is more significant. It is the difference between a hiker completing a single step and the hiker completing segments of a path that required numerous steps.

In this section we discuss the activities and outcomes that are present in product development and how to sequence them. We'll also introduce a visual representation of the sequence called an *activity map*.

Activities and Outcomes

Design activities and outcomes are intimately connected, and either can be considered primary. That is, we can say that design evolution occurs through a set of intentional activities whose results are outcomes. In this statement, the activities are the essential elements of development, with the outcomes being ancillary results of the activities.

Alternatively, we can say that design evolution consists of achieving a desired set of outcomes. In this view, the outcomes are the essential elements of development, with the ancillary activities being used to create the outcomes.

Both views can lead to successful product development. Both views can be found in common project scheduling theory and tools. The selected viewpoint is largely a matter of personal choice.

In this book, we choose to consider the outcomes as the primary elements in product evolution. This is because the completion of a stage of development depends on the availability of certain transferable product development artifacts[1], rather than the completion of certain activities.

In addition, there may be multiple ways of achieving a particular outcome. When such is the case, we believe it is better to specify the outcome and provide the development

---

[1]Product development artifacts are objects made by the product development team; examples include sketches, test reports, requirements tables, and so on.

team the flexibility to choose the activities they will use to achieve the outcome.

Thus, we recommend that product evolution focus on artifacts that are transferable representations of design outcomes. Such artifacts can be used both to evaluate the current state of evolution and to support future evolution. These artifacts are more useful when they are concrete, specific, and unambiguous.

When an outcome is captured in a product development artifact, good design practice requires critical review of these artifacts. The team must assess the quality of the design decisions made, and should iterate through the activity until the outcome is demonstrated to be desirable. The process of creating artifacts and critically evaluating their quality is essential to high-quality product development.

Activity Maps

Whether explicit or not, every project has an underlying network of design outcomes that defines the relationship between multiple design outcomes and the design activities that lead to them. We call these *activity maps*.

Activity maps indicate the relationships between design activities and outcomes, whether they be independent, dependent, or interdependent. These relationships help the team decide the order in which outcomes should be pursued and consequently choose the order of design activities. This important information dictates, to a large degree, how the efforts of the different people on the product development team should be coordinated.

When considered in a detailed (low-level) way, activity maps are different for every project. When considered at a less detailed level, activity maps are similar or identical from project to project.

An activity map is a good way to visualize the structure of activities and outcomes during product devlopment. A fairly high-level activity map for developing the requirements for a human-powered water well drill for the developing world (see AppendixB) is provided in Figure 2.5 as an example.

Activity maps use nodes to represent design outcomes, arrows to represent design activities, and double-headed arrows to represent interdependent relationships, as shown in Figure 2.6. In activity maps, the symbols are arranged to show the logical sequencing of activities and outcomes.

Arrows leaving a node indicate that the activity cannot be started until the outcome represented by the node is available. Arrows pointing to a node indicate that the activity represented by the arrow is necessary to create the outcome represented by the node.

Multiple activities can depend on a single outcome, and multiple activities can be required to achieve an outcome. But each activity on the map leads to one and only one outcome.

Multiple outcomes can be combined into a compound outcome as shown in Figure 2.6. When multiple outcomes are needed as input for a design activity, it is convenient and clear to represent them as a single compound outcome.

To more fully understand this, consider Figure 2.5. Looking at only the two leftmost arrows, we can see that Outcome 1 is achieved as a result of interviewing the client and that creating and revising the project objective statement cannot begin until the client interview is complete. The Project Objective Statement (Outcome 2) results from the activity "Craft and revise project objective statement."

Once Outcome 2 is achieved, the activities of interviewing well drillers, searching the internet, and benchmarking competitive products can begin in any order or simultaneously. The outcome of each of these activities (3, 4, and 5) is shown as being interdependent on the outcomes of the others. This is represented by the dashed arrows in the figure. What this means is that the interviews with the well driller will influence what internet research to do and vice versa. In such cases it is likely that multiple interviews or multiple internet searches will need to be completed before a mutually acceptable outcome is achieved for both activities. The interviews

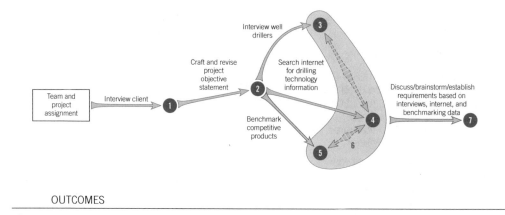

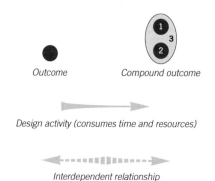

Figure 2.5: Top-level activity map for developing initial requirements for a human-powered water well drill for the developing world.

Outcome

Compound outcome

Design activity (consumes time and resources)

Interdependent relationship
(consumes no time or resources)

Figure 2.6: Activity map symbols. The compound Outcome 3 is achieved by having both Outcomes 1 and 2.

and the internet research are also influenced by benchmarking competitive products. All three of these activities are interdependent. As discussed shortly, interdependent activities require careful coordination.

Importantly, establishing requirements can only begin once the compound Outcome 6 has been completed. This means that

"Discuss/brainstorm/establish requirements" depends on all three prior activities.

* * * * *

While developing your product, you will encounter times when the amount and diversity of activities that needs to be done will be significant. At those moments it can be quite useful to map out the relationship between key outcomes and activities. Even when this is done in an informal way (with a pencil and scratch paper), it will help you clearly identify the most critical things to focus on.

It is equally important to know, however, that you won't need to map out everything. There may be times when an explicit activity map won't help you accomplish your goal any more efficiently or effectively. Judging when and when not to use an activity map is exemplified by the hiker who does not chart a detailed path across a meadow, but finds she must before scrambling up a rock face. Just like her, you'll need a mapping when the stakes are high and the way forward requires careful treading.

Nevertheless, whether it is explicitly drawn or not, there *is* an activity map that describes your work.

To help you through those moments where explicitly mapping out the path is beneficial, let us add some formality to the above discussions. Design activities and outcomes have one of only three possible relationships with each other: an independent relationship, a dependent relationship, or an interdependent relationship. These relationships are shown as activity maps and are described in Figures 2.7 and 2.8.

Independent Activities

Independent activities have no impact or influence on each other. In relation to each other, it does not matter which design activity is accomplished first.

Dependent Activities

Dependent activities must be carried out in series, as shown in Figure 2.8. The defining characteristic of dependent activities is that the outcome of one is the starting point to another activity.

Interdependent Activities

The last way that multiple design activities can be arranged is also shown in Figure 2.8. Interdependence is indicated by the dashed arrow. Here, the preliminary outcome of one activity affects another parallel activity. One characteristic of these interdependent activities is that they can be serialized *and* either activity could be carried out first. Once serialized, however, the activity executed first may be re-executed based on new information gained in the second activity. This process can (and should) be repeated until a mutually acceptable outcome for both activities is converged upon. Activity maps don't explicitly show open-ended iteration, but wherever there are interdependent activities iteration will likely be required.

2.4   Coordination of Effort

Virtually every product development effort includes more than one person. For maximum benefit, the efforts of all individuals must be coordinated to best meet the needs of the product. With poor coordination, effort will be duplicated and time will be wasted. In this section we discuss the impact of the three dependency relationships (independent, dependent, and interdependent) on project coordination.

Coordinating Independent Activities

Independent activities should be carried out in parallel whenever possible to reduce the overall product development time. Because the activities and their outcomes are independent, there are no logic-related barriers to parallelizing them.

Coordinating Dependent Activities

Dependent activities must be carried out in series. The critical path, or the set of activities that determines the overall product development time, will involve multiple dependent activities (See Critical Path Analysis (11.13) of the Development Reference). To that end, the product development team should pay close attention to dependent activities and make sure their progress is well-tracked since failure to successfully complete them affects downstream product development efforts. Whenever it is possible to decouple dependent activities and make them independent, it is generally worth doing.

Coordinating Interdependent Activities

Interdependent activities require the most coordination, since the outcome of one activity affects another parallel activity. By their nature, they will need to be carried out in parallel. However, the outcome of any one of the parallel activities cannot be finalized until the outcomes of all the other parallel activities are shown to be satisfactory. Individuals or subteams working on interdependent activities need to plan on having a structured scheme for collaboration. Perhaps the easiest way to do this is for the team to establish a short interval between subteam coordination, where the interval is no longer than the amount of lost time the subteams are willing to live with.

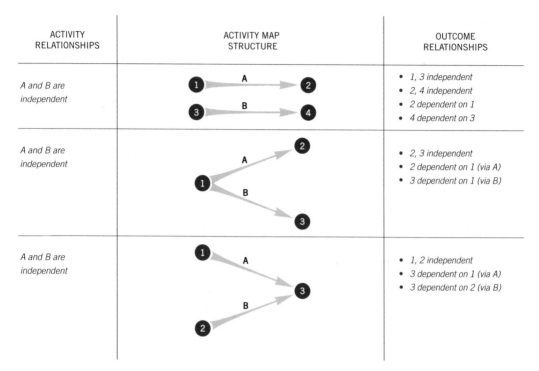

ACTIVITY RELATIONSHIPS	ACTIVITY MAP STRUCTURE	OUTCOME RELATIONSHIPS
A and B are independent		• 1, 3 independent • 2, 4 independent • 2 dependent on 1 • 4 dependent on 3
A and B are independent		• 2, 3 independent • 2 dependent on 1 (via A) • 3 dependent on 1 (via B)
A and B are independent		• 1, 2 independent • 3 dependent on 1 (via A) • 3 dependent on 2 (via B)

Figure 2.7: Independent activities and their activity map logic.

Activity maps and other coordination tools can be constructed with various levels of detail, as shown in Figure 2.9. At the highest possible level, the product development process can be described with a single outcome and a single activity. While this very high level of detail is not useful, it's worth noting that this activity map applies to *all* manufactured products, and that this is the only activity map that applies to all manufactured products.

Every level of increasing detail represents a narrowing of that map's applicability to other products. For example, at the second level down (as shown in the figure), we find a generic set of activities that can be used to develop an *engineered* and manufactured product. Most of the generic product development processes in the engineering literature exist at this level of abstraction.

While some, both experienced and novice, will find this minimal level of detail insightful, others need a greater level of

detail in order to be successful or to facilitate coordination between product development team members. At their greatest level of detail, each activity map is unique and can ultimately be used to create a custom project plan for a specific project. Creating customized project plans is discussed in Chapter 8.

Importantly, you'll need to ask yourself *how much detail is enough to facilitate coordination among product development team members?* To help answer that question for your particular situation, consider these principles:

• The activity map does not equal coordination; it is only a tool to facilitate good coordination. Therefore, the activity map alone serves no purpose until it is used by the team.

• A person working alone (perhaps on an independent activity) need not decompose that activity into

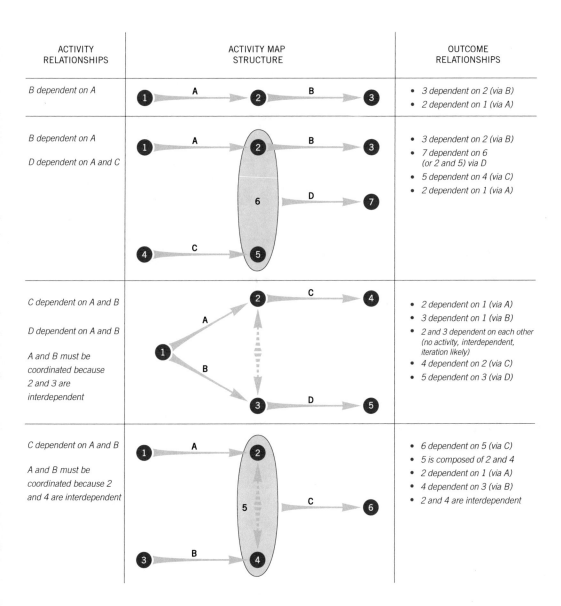

Figure 2.8: Dependent activities and their activity map logic.

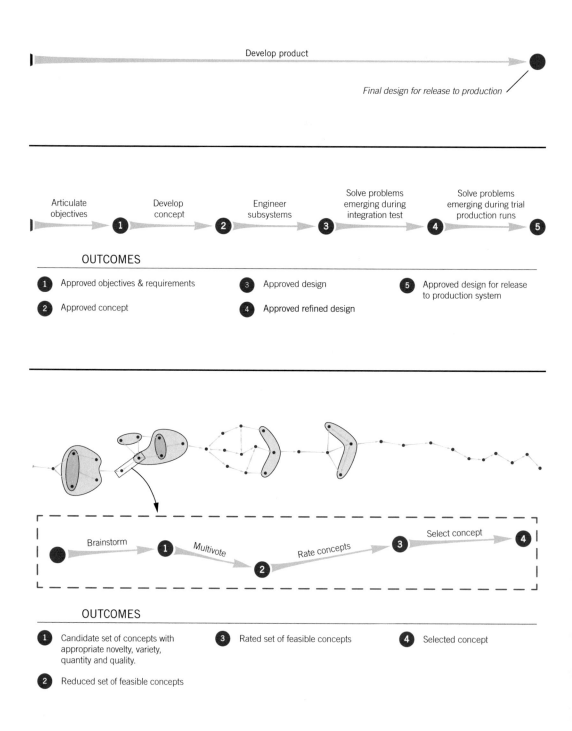

Figure 2.9: Activity map levels. The top map shows the highest level process, while the bottom map shows the lowest level outcomes and activities. Intermediate maps show intermediate levels of detail.

sub-activities to facilitate coordination. Such a designer may, however, find decomposition to be a valuable way to divide a complex task into simpler dependent tasks.

- Most product development teams have a *coordination interval* of sorts, which represents how often the team meets. Activities done by an individual alone can be decomposed to the level of one activity per coordination interval. Activities done by subteams can be further decomposed into activities that can be done by an individual within the coordination interval for the subteam (which will be shorter than the interval for the team as a whole).

2.5 Stages of Development

Up to this point in the chapter we have discussed two main concepts: small-scale evolution, and medium-scale evolution. We spoke of small-scale evolution when we introduced the STEP cycle and compared it to footsteps in a journey across a terrain. We spoke of medium-scale evolution when we introduced design activities and activity maps and compared them to segments of a path, which require multiple footsteps to traverse.

We now discuss evolution on a large scale by introducing the stages of product development and comparing them to the major milestones in a large journey.

Designs generally evolve through distinct stages of development, where each stage has clearly defined characteristics. Throughout each stage, the design is evaluated to determine how well it is evolving. Each stage culminates in an approval that allows the next stage to begin with minimal risk. Understanding these stages will help you establish an overall project goal for a specific project, help you define the starting and ending points for the project, and give you a sense for how resources will need to come together to evolve the design through the pertinent stages of product development.

For all manufactured engineered products comprising more than one part, there are six stages of product development. The six stages are shown in Figure 2.10. They are:

1. Opportunity development

2. Concept development

3. Subsystem engineering

4. System refinement

5. Producibility refinement

6. Post-release refinement

The stages of product development are fundamentally stages of design evolution, where each stage is named for the primary focus of the evolution during that stage. For example, during the subsystem engineering stage, the focus is on creating a transferable definition that has been demonstrated to be desirable for each engineered subsystem.

While the design evolves through the stages of development, two other things are evolving simultaneously and in an interrelated way. They are the requirements and the tests, which are employed as a way of verifying whether or not the current design is desirable. Figure 2.10 illustrates this.

Prototypes and models play a significant role in helping the team evolve a high-quality design. As shown in Figure 2.10, prototypes and models are constructed based on the current design, and are used in tests to estimate how well the product will meet the requirements. Note that for visual simplicity, we have shown only one model and prototype icon per stage. But any number of prototypes and models can (and probably should) be created for each stage of development.

When the design has evolved to the point that it has been demonstrated to meet the purpose for the stage, approval is obtained.

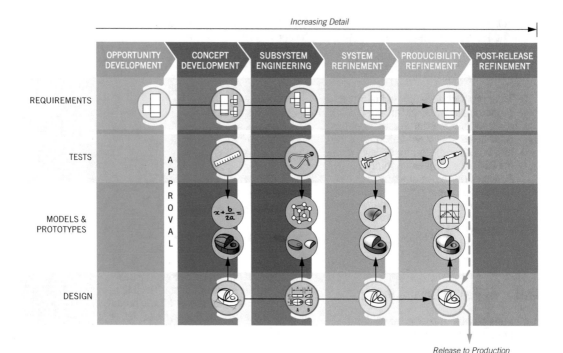

Figure 2.10: The requirements, tests, and design all evolve through the stages of product development; the design is eventually released to the production system for manufacturing. Prototypes and models are instantaneous representations of the design that are used in tests to determine predicted and measured performance that can be compared to the requirements.

The approvers generally include the client[2] as well as people from the development team's organization that are external to the team.

Note that at the end of the third stage, every aspect of the product has been engineered. However, this does not mean that the engineering on the product is complete. It continues in the last three stages to refine the design as the product moves closer to (and beyond) release to the production system. Also note that at the end of the

producibility refinement stage, the design defines a product that is market-ready. This is when the design is released to a production system for mass production.

There are two subtle, but important, things to glean from Figure 2.10. The first is that the six stages of development are — in reality, not just in the figure — generally sequential. We describe them as sequential because it is highly desirable, and quite possible, that the design continuously evolves to higher and higher states of definition and desirability. Careful transition between stages ensures this. While these transitions are discussed in more detail in Section 2.6, it is sufficient for the present discussion to say that the completion of each stage consists of obtaining approval of the design. Once a

[2]The client is the person or organization that provides the resources for completing the development project and therefore has a significant stake in the outcome. Clients are rarely part of the development team. But when they are, the approval needs to be an external action to maintain objectivity.

design has been approved as having completed a stage, it is undesirable to have the design regress to a previously completed stage. Such devolution, as it were, is costly and emotionally painful for the product development team.

The second subtlety to glean from Figure 2.10 is the controlled overlap of adjacent stages of development. The chevron shapes shown at the top of the figure are used to convey this. While there are legitimate reasons to carry out two adjacent stages of development concurrently — such as to keep the development team actively evolving the design while the approvers are evaluating the design at the end of a stage — it is essential to recognize that there is risk in the overlap.

What is the risk? The risk is wasted development effort. Consider, for example, the overlap that may exist at the end of the concept development stage and the beginning of the subsystem engineering stage. If a team decides to begin engineering parts (which is fundamentally part of the subsystem engineering stage) before a final concept has been approved (which is fundamentally part of the concept development stage), the team risks wasting resources used to engineer subsystems that may never be used. Clearly, the risk of wasted development should be minimized, but minimizing risk must be balanced against other pertinent concerns facing the team.

Activity Maps for Product Development

In Chapters 4–7 we present top-level activity maps for each stage of product development. To effectively manage the product development project, however, customized more detailed maps will need to be created by the team for each stage of development. Nevertheless, the top-level maps provided are helpful starting points for each stage.

Figure 2.11 shows a top-level activity map for the concept development stage. Even though this map is for a single stage – and is only top-level in detail – it shows that

product development is not trivial nor easy. The activities and outcomes that need to be accomplished, and their coordination, take a significant amount of work. If thoughtfully coordinated, however, the product development journey can be smoother, resulting in a more enriching experience for the development team and a better product for the client.

While it might be interesting to visualize the whole activity map for product development (meaning maps from all stages of development together), this map would be too complicated to show on a single page. For this reason, we break the process down into manageable stages and discuss each in detail (Chapters 4–7) and provide activity maps on a stage-by-stage basis. This approach has the advantage of allowing the team to focus on one stage at a time.

2.6 Approval at Stage Completion

To reduce the risk of wasted effort during product development, the design's desirability and transferability should be formally evaluated and approved at the end of each stage of development. This section describes the kinds of tests required for approval.

Approval is generally based on the successful completion of (i) artifact checks, (ii) performance tests, and (iii) validation tests. The relationship between these is illustrated in Figure 2.12.

Artifact checks are performed to ensure that the requirements, the tests, and the design — as captured in transferable artifacts — are clear and complete, and match the product development team's intent. Artifact checks should be performed by a member of the development team other than the person who created the artifact. To avoid mistakes, a formal process for checking product development artifacts should be implemented and followed. The design should be checked to see if any information is missing or any ambiguities exist.

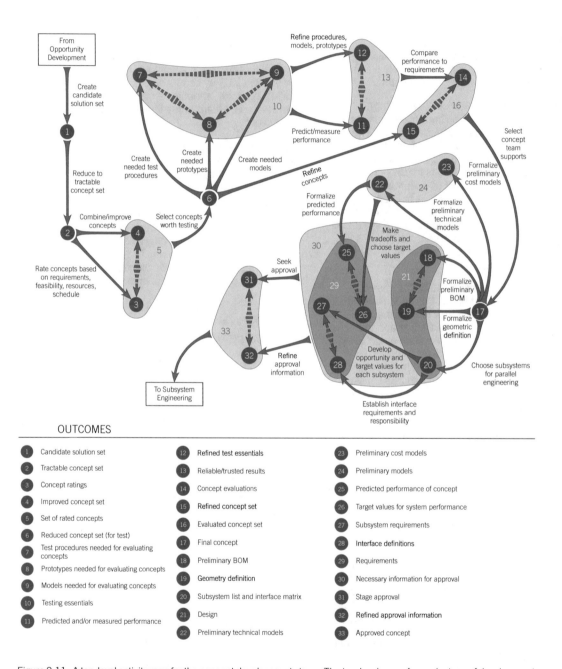

OUTCOMES

① Candidate solution set	⑫ **Refined test essentials**	㉓ Preliminary cost models
② Tractable concept set	⑬ Reliable/trusted results	㉔ Preliminary models
③ Concept ratings	⑭ Concept evaluations	㉕ Predicted performance of concept
④ Improved concept set	⑮ **Refined concept set**	㉖ Target values for system performance
⑤ Set of rated concepts	⑯ Evaluated concept set	㉗ Subsystem requirements
⑥ Reduced concept set (for test)	⑰ Final concept	㉘ **Interface definitions**
⑦ Test procedures needed for evaluating concepts	⑱ Preliminary BOM	㉙ Requirements
⑧ Prototypes needed for evaluating concepts	⑲ **Geometry definition**	㉚ Necessary information for approval
⑨ Models needed for evaluating concepts	⑳ Subsystem list and interface matrix	㉛ Stage approval
⑩ Testing essentials	㉑ Design	㉜ **Refined approval information**
⑪ Predicted and/or measured performance	㉒ Preliminary technical models	㉝ Approved concept

Figure 2.11: A top-level activity map for the concept development stage. The top-level maps for each stage of development are shown in Chapters 4–7. The overall map for product development could be created by combining the individual maps for each stage sequentially, but this map would be too complicated to show on a single page.

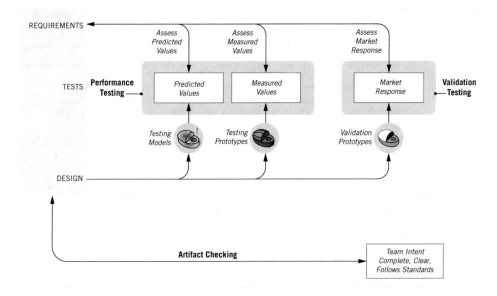

Figure 2.12: Evaluation of desirability and transferability testing during a development stage. This shows the relationship between artifact checks, performance tests, and validation tests. The design is checked for quality and completeness. Using prototypes and models based on the design, product performance is measured or predicted using the tests. Validation testing consists of having a market representative evaluate a prototype and pass judgment on the quality of the design. For approval, all required information should be present, the predicted and/or measured performance should meet the ideal or target values, and the market response should demonstrate that the market requirements are met.

A two-level artifact check should be applied during the checking process. First, the design should be reviewed to determine whether someone trained in the art (but not the specific product) could reproduce the product from the design without any outside help. Second, the design should be reviewed to determine whether any loopholes exist in the design that could allow a low-quality product to be manufactured.

Performance tests are carried out to determine predicted and/or measured values of the requirements. Models and/or prototypes are made from the current design and evaluated using the current test procedures. Performance tests on models create predicted values; tests on prototypes create measured values. Before final approval of a stage, the predicted and/or measured values must meet or exceed all the acceptable limits, and as many of the target values as possible. Acceptable limits and target values are part of the requirements as discussed in Chapter 4.

Performance tests are carried out by members of the product development team.

Validation tests are managed by the product development team, using market representatives[3] to provide the evaluation.

[3]A market representative is someone or some group of people chosen to represent the market. A representative can be thought of as a customer, an end user, etc. Because validation is fundamentally about outside approval, it is important that the market representatives be chosen from outside the product development team.

CHAPTER 2. PRODUCT DEVELOPMENT FUNDAMENTALS

Table 2.1: Overview of desirability and transferability testing. This table summarizes the artifact checks, performance tests, and validation tests that are performed during each stage of product development.

	Artifact Checks	Performance Tests	Validation Tests
Test objective:	Check the quality and completeness of the design, requirements, and tests. Ensure that these fully and unambiguously capture what the team intended.	Use models and prototypes based on the design to predict and measure the performance of the design according to the tests. Compare the predicted and measured performance with the requirements.	Have the market representatives evaluate design artifacts to evaluate the quality of the design from the market perspective.
Artifacts used in testing:	Transferable medium appropriate for communicating the design (e.g., written list, technical drawing, bill of materials, test report, requirements matrix)	Models or prototypes based on the design; standard test methods that govern testing	Prototypes that communicate appropriately to the market representatives; standard test methods that govern validation tests
Artifacts created during testing:	Formal approval of the artifacts checked; usually includes release of a given revision	Test reports indicating the results of applying tests to models and prototypes; predicted and/or measured values (in a requirements matrix, for example)	Test reports indicating the results of validation tests; market response values (in a requirements matrix, for example)
Who is involved:	Evaluation performed by member(s) of the product development team, excluding the person who created the artifact	Evaluation performed by product development team	Evaluation coordinated by development team and performed by market representatives

Validation tests require prototypes that can be evaluated by the market representative. The goal of validation tests is to determine whether the product as designed will satisfy the market. Because the team is *not* the market, this judgment must not be made by the team.

This notion is captured in Figure 2.12, where the market requirements are shown to be validated principally against the validation prototype, not the design. Such validation is essential, for it is not possible that the product development team can know all the intentions of the market. For this reason, the market representative becomes an essential part of the validation tests.

Table 2.1 summarizes the nature of artifact checks, performance tests, and validation tests.

When seeking approval, the test results must be clearly conveyed to the project approvers. This is rarely trivial since the approvers are generally not part of the product development team and may be from diverse disciplines such as marketing, engineering, production, sales, support, service, and finance. Approval is not complete until all project approvers have approved the design.

A top-level summary of the process of obtaining approval is provided in Section 2.7.

2.7 Obtaining Design Approval

Design approval must be obtained at the end of each stage. However, partial design approval may be obtained at intermediate points in each stage, whenever an important piece of the design is complete.

Conceptually, obtaining design approval is relatively straightforward — simply demonstrate to the project approvers that all of the required elements of the design meet the requirements. However, because the requirements and the design are co-evolving, and because the choice of which tests to carry out significantly affects the approval, the activity map is somewhat complex.

Starting with the requirements, tests, and design that are up for approval, Figure 2.13 shows a general activity map for obtaining approval. The activities shown in the map include the following items.

Check Product Development Artifacts

The first activities in obtaining approval are checking of the requirements, tests, and design. The product development artifacts should be formally checked to ensure they meet appropriate standards for the current stage of development, as summarized in Chapters 4–7. When artifacts have passed the check, it's conventional and wise for them to be annotated with a date and the checker's name or initials.

Prepare Test Models and Prototypes

When the design has passed the checks, the artifacts to be used for testing are prepared. These include models and prototypes. Each of these artifacts should be based on the design that has passed the checks. The specific artifacts to be created are determined by the performance tests and the validation requirements.

Test the Performance

The prototypes and models are tested according to the checked test methods. It is important to follow these methods to obtain repeatable and consistent results.

The results obtained from testing are recorded in transferable form as part of the test information. The performance test results must also be checked as part of the product devlepment artifacts, and should be assigned a date and a revision number.

Evaluate Test Results

The results obtained from performance testing are compared with the requirements. An evaluation is made as to how well the predicted or measured performance matches the desired values. This evaluation is recorded in a transferable artifact that will be reviewed by the project approvers.

Perform Validation Testing

The validation prototypes are shared with the market representatives and the assessment of the market representatives concerning the desirability of the product is obtained.

Convey Results for Approval

The checked design, the evaluation of the performance testing, and the market validation results are conveyed to the project approvers for approval. The approvers review the items conveyed and make a decision whether or not to approve. If approval is granted, the design is approved and the team moves fully into the next stage of development. If not, changes are made to the design to resolve weaknesses, and any necessary steps to obtain approval are repeated.

2.8 Summary

This chapter has focused on the concept of design evolution – the incremental and iterative transition from a product idea to a manufactured product on the shelves of a store. We discussed that transition on three scales, and compared it to a journey by foot over terrain.

The smallest scale, which was compared to footsteps, is the STEP cycle. This cycle underlies everything that is done during

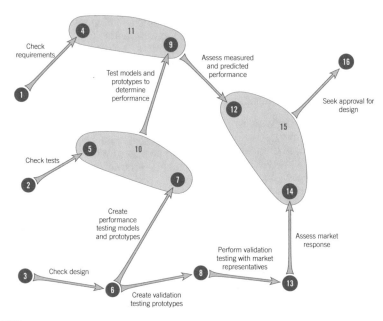

OUTCOMES

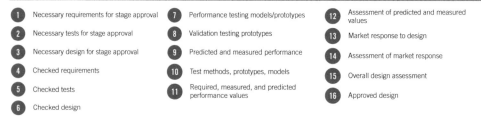

1 Necessary requirements for stage approval	7 Performance testing models/prototypes	12 Assessment of predicted and measured values
2 Necessary tests for stage approval	8 Validation testing prototypes	13 Market response to design
3 Necessary design for stage approval	9 Predicted and measured performance	14 Assessment of market response
4 Checked requirements	10 Test methods, prototypes, models	15 Overall design assessment
5 Checked tests	11 Required, measured, and predicted performance values	16 Approved design
6 Checked design		

Figure 2.13: Generic activity map for obtaining approval for a design.

product development; it is the main engine of evolutionary progress. While it may help to think about the four parts of the cycle explicitly — especially as you train yourself in its use — those steps are mostly executed in a natural and informal way, just as the hiker's footsteps are natural and informal.

The middle scale was compared to segments of a chosen path over the terrain. To the designer, these segments are activity maps. There are times during product development when the activity map is simple, and others when it is complex. Charting a path through those complexities

can be greatly facilitated through the construction of activity maps.

The largest scale discussed in this chapter was compared to major milestone in the journey. These were described as the stages of product development. Because it can be so easy for a product development team to get off course, it's essential to evaluate the development progress at major milestones, and make course corrections before continuing on. The stages presented in this chapter exist for all manufactured and engineered products. For this reason, it's extremely valuable to understand the characteristics of those stages since we'll face them again and again throughout our careers.

A good portion of this book is dedicated to characterizing the stages of development and providing you with a top-level activity map for evolving the design through each stage. We recognize, however, that lower-level activity maps (the level at which day-to-day work is actually done) cannot be generically specified for all products. Therefore the opportunity to plan the journey is also yours. Chapter 8 is dedicated fully to helping you do this for any setting.

At this point we recognize that these fundamentals can be overwhelming and that it will take time and effort for you to absorb, practice, and master them. Nevertheless, developing a strong understanding of these fundamentals will help you become an effective product developer. As you work with these fundamentals, they will become a natural part of your approach to product development.

2.9 Exercises

Test Your Knowledge

T2-1 What are the three kinds of product development artifacts that evolve during the product development process?

T2-2 What are the three characteristics of useful outcomes in activity maps?

T2-3 What are the two kinds of product development artifacts that are used to determine how well the design meets the requirements?

T2-4 What are the four parts of the STEP cycle?

T2-5 What are the six stages of product development?

T2-6 What are the three different evaluation methods used to obtain approval?

T2-7 For stages where performance can be predicted or measured, what level of performance is needed to achieve approval?

T2-8 What is a market representative?

Apply Your Understanding

A2-1 In your own words, why is design evolution largely iterative and incremental?

A2-2 Explicitly follow the STEP cycle as you create a solution to a problem (perhaps try this for an engineering analysis problem, a writing problem, a design problem, or any other problem that requires the creation of some specific artifact). For each step, explain the actions you have taken.

A2-3 In your own words, explain the meaning of independent activities, dependent activities, and interdependent activities.

A2-4 Prepare a thoughtful paragraph about the risks of not coordinating team member efforts during product development. Be precise about the risks that will be faced. Write the paragraph as if it will be used to train new designers.

A2-5 Quiz yourself on activity map logic, by covering the activity relationships and the outcome relationships of Figures 2.7 and 2.8, and stating them based on the map structure.

A2-6 Activity maps can be used for many activities outside of product development. Demonstrate your understanding of activity maps by creating one for purchasing a new tablet computer. Identify independent, dependent, and interdependent activities and outcomes.

A2-7 Choose a relatively complex task with which you are familiar and develop an activity map for accomplishing the task. In your activity map,

Required product development artifacts for each stage of development. Use with Exercise A2-12.

	Opportunity Development	Concept Development	Subsystem Engineering	System Refinement	Producibility Refinement	Post-release Refinement
Requirements						
Tests						
Design						

demonstrate your understanding of independent, dependent, and interdependent relationships by having at least one of each. Also demonstrate your understanding of compound outcomes by having at least one.

A2-8 Consider a multi-person project you are currently working on. Give yourself an honest evaluation of how well you are coordinating the efforts of the people involved, or how well you are following a project plan someone else has created that involves you. If you don't know how to evaluate yourself in this area, find a colleague or mentor who help you through the evaluation.

A2-9 Through which stages of development will your design need to pass for your specific project?

A2-10 What is the role of approval in evolving the design?

A2-11 What role do market representatives play in evolving the design?

A2-12 Complete the table here using your own words to indicate the required product development artifacts at the end of each stage of development.

A2-13 Consider a project you're working on now.

    a) Who is the market representative?

    b) In what ways has your team used market representatives to evolve the design information?

    c) What risks are associated with your use of market representatives, and how should you mitigate those risks?

CHAPTER **3**

Design Skills

The evolution that takes place during product development, as the design transitions from an embryonic idea to a complete design for a manufactured product, is not automatic nor is it a mechanical process. It does not occur on its own or by chance. Importantly, excellent products are not created by rote application of specified design methods. Instead, they are created when designers apply their best and most skillful efforts to the challenge at hand.

The purpose of this chapter is to explain ten crucial design skills that we need to apply when developing great products. These skills are:

- Planning
- Discovering
- Creating
- Representing
- Modeling
- Prototyping
- Experimenting
- Evaluating
- Deciding
- Conveying

We explain these ten design skills and categorize them as crucial, because either in an individual setting or in a group setting, these skills *will be needed* to evolve the design through the stages of development. In fact, they are often used during every stage of development and across every activity map. This means that *discovering*, for example, is a skill the designer will use more than once and during more than one stage of development, or across more than one activity map. It will be something the designer does multiple times, even continuously in some cases.

Not only will these ten skills be needed to cause design evolution to occur, they will also be used to judge how good we are as designers. Understandably, others will evaluate our skills in these areas before we're assigned to a new product team, or before we win a bid for a job, or before we gain employment. As such, these skills are crucial to our profession and deserve our ongoing attention.

In many cases, design tools and methods have been developed to help us acquire and refine these crucial design skills. Consistent, repetitive practice with these tools and methods leads to the development of greater skills. Many of these tools and methods are included in the Development Reference, which is Part II of this book.

The ten design skills are briefly presented below with referral to the Development Reference for more information. As you read through the skills, take heart in knowing that we can *all* improve in these ten areas.

3.1 Planning

Planning is the act of considering why a design effort is needed, what resources enable it and what constraints limit it, *and* allowing that information to guide our actions to be effective at achieving the desired outcome and be efficient in using resources.

Planning, as a skill, is needed continuously throughout the product development process. It is needed formally, and informally. For large complex projects, for simple ones, and for small pieces of projects. Planning is needed for people working alone and for people working in teams. Planning keeps us from wandering.

Designers who have developed the skill of planning are curious about the following, and do what they can to get and use this information:

1. The purpose and objective of a project/task

2. The starting and ending point of the project/task

3. The constraints of the project/task

4. The resources available to carry out the project/task

5. The schedule related to the project/task.

At any point while working on a project, it is valuable to check yourself to see if you know the information listed above, and if you're using it to improve effectiveness and efficiency.

Specific activities described in the Development Reference related to planning include: Critical Path Analysis (11.13),

Goal Pyramid (11.32), Nucor's Circles (11.38), Objective Tree (11.39), Plan Do Check Act (Shewhart Cycle) (11.46), Planning Canvas (11.47), Project Objective Statement (11.49), and Value Engineering (11.69).

Understanding and practicing these activities will help you develop skills in planning.

3.2 Discovering

Discovering is the skill of finding existing knowledge and ideas. It also includes the skill of uncovering customer needs and preferences. Designers who are excellent at discovering are better able to build on previous work and develop a desirable product than those who are not. It's virtually always faster to find knowledge than to create it. A colleague of ours has expressed this concept by saying "I can save myself hours in the library by spending months in the laboratory." In the time-critical world of product development, we cannot afford to overlook existing information that is relevant to our design.

People who are skilled at discovering recognize that at the beginning of a project they do not know enough to be successful. To gain the needed knowledge, they are persistent and creative. They know that relevant information exists somewhere, and they are relentless in their pursuit of that information. When they build on existing information, they "stand on the shoulders of giants," as stated by Isaac Newton.

Specific activities that can be used to develop your discovering skill are treated in the Development Reference. They include Catalog Search (11.7), Codes and Standards (11.8), Delphi Method (11.15), and Internet Research (11.33).

Specific activities that can be used to help uncover market needs and desires are also covered in the Development Reference. Including Focus Groups (11.31), Interviews (11.34), Observational Studies (11.40), and Surveys (11.65).

The better you are at discovering, the more likely you will be able to create a world-class product.

3.3 Creating

Creating is the skill of generating new knowledge or combining old knowledge in new, unexpected ways. Key actions that are part of this skill include: organizing, imagining, formulating, brainstorming, synthesizing, substituting, and conceiving.

Inexperienced designers may think that creating is the primary skill needed in product development. Certainly creating is an important skill. However, if creating is not coupled with other design skills, the full value of the creations is never fully realized.

Skill in creating requires the ability to suspend judgment as well as the ability to apply judgment. If judgment is not suspended early in the creation process, promising ideas will be rejected as poor, before they have had the chance to be revised and improved. If judgment is not applied later in the creation process, the designer will be unable to identify weaknesses of the creation, which will lead to inferior designs.

There are many methods that have been successfully used to enhance the creative process. The Development Reference includes summaries of Bio-Inspired Design (11.4), Brainstorming (11.5), Concept Classification Tree (11.9), Decomposition (11.14), Method 635 (11.35), Mind Maps (11.36), Recombination Table (11.53), SCAMPER (11.58), Storyboards (11.64), and Theory of Inventive Problem Solving (TRIZ) (11.66), all of which can help develop the skill of creating.

3.4 Representing

Representing is making graphical representations of the product to communicate the design to others (physical representations of the product are considered prototypes and are covered in Section 3.6). Graphical representations are used to explain the idea of the design to others, to define the geometry and other physical characteristics of the product, and to promote the product to those who will approve it, among other uses.

At least four types of graphical representations are used.

Sketching

Sketches are used as quick graphical representations of a product. Sketches can be simple 2-D representations with annotations to help explain the sketch. They can be more highly polished 2-D representations. They can be perspective or isometric sketches that give an appearance of 3-D. The best sketch to use will depend on the purpose for the sketch.

Sketching is a skill that needs work to be developed. The Development Reference entry on Sketching (11.63) contains some simple suggestions to help you develop your sketching skills.

Effective sketching is a great help in communicating the intent of your design both within and without the team.

Engineering Drawing

Engineering drawings are drawings that are made accurately and precisely, using drawing tools or precise CAD drawing packages. Engineering drawings are made to scale and will comprise a major part of the design.

Engineering drawings can be 2-D representations or isometric representations that communicate some 3-D information. Most engineering drawings will have multiple views, each of which contains 2-D information about that particular view of the 3-D object.

Engineering drawings can be part drawings or assembly drawings.

Engineering drawings can be used early in the product development process. As the design evolves, the engineering drawings will generally increase in detail.

Standard practices should be followed when creating engineering drawings.

The Development Reference entry on Drawings (11.23) contains more information on engineering drawings. The entry on Drawing Checking (11.22) describes how a drawing is checked for quality.

Solid Modeling

An effective graphical representation tool for geometry information is solid modeling. In solid modeling, a virtual model is created that has identical geometry to the eventual physical part. Solid model creation can often follow some of the same steps as a physical machining process.

Solid models can be directly used for analysis of volume, mass, and center of mass. Meshes can be created from the solid model to support finite element modeling. STL files can be created from solid models for use with additive rapid prototyping machines. CNC tool paths can be created from solid models for use in machining the modeled parts. Engineering drawings can be created from solid models.

While solid modeling is not essential to product development, most companies doing product development use solid modeling as part of their development process.

Solid modeling is further described in the Development Reference entry on CAD Modeling (11.6).

Presentation Graphics

Presentation graphics are graphical representations that are prepared to communicate the design to people outside the development team, generally to a non-technical audience. They may include sketches that have been colored with chalk, marker, or airbrush; computer-generated renderings from solid or surface models; photographs of rough physical prototypes; or other high-quality graphical representations.

Presentation graphics can have a great effect on the audience. A product that has high-quality presentation graphics is often perceived as better than one with moderate- or low-quality graphics. As you seek to convey your work, you will be wise to make sure your presentation graphics reflect the quality of your design.

3.5 Modeling

An essential characteristic of engineering design is the use of models to predict the performance of the design. The skill of modeling allows the team to make rapid iterations in changing design parameters to achieve the desired performance.

Engineering models may be analytical, numerical, or statistical. The common thread behind engineering models is that they allow the designer to make predictions about how real systems will perform. When engineering models have been validated to match reality, they allow rapid design through exercising models, rather than just building a series of prototypes.

When thinking of creating engineering models, development teams should consider the creation of multiple models, rather than a single model. Early models are likely to be coarse, while later models will be more refined. Both types of models are useful in product development.

Section 6.5 provides a description of how modeling can be used in product development. The Development Reference includes entries on Design of Experiments (11.18), Dimensional Analysis (11.21), Finite Element Modeling (11.30), Sensitivity Analysis (11.61), and Uncertainty Analysis (11.68), all of which are related to the skill of modeling.

3.6 Prototyping

Prototyping is the skill of creating physical approximations to the product based on the current design that will help answer important questions about how well the product works. Key actions that are part of this skill include mocking up, machining, casting, 3D printing, and breadboarding. Designers who are skilled in making quick, effective prototypes are highly valued.

A detailed discussion of prototyping — including the six types of prototypes — is found in the Development Reference entries for Prototyping (11.50) and Rapid Prototyping (11.52).

3.7 Experimenting

Experimenting is the skill of using models, prototypes, and people to measure how the product should and does work. Key actions that are part of this skill include setting up experiments, instrumenting test setups, statistical analysis, and working with human subjects.

Formal and informal experimentation is extremely valuable to the goals of product development. Informal experimentation involves simply *trying it*. We often do this by spending very little money and time setting up an experiment. The objective with such experiments is to very quickly get a sense for if something is likely to work or not. A natural consequence of this is that, unlike formal experimentation, the results are generally anecdotal.

Formal experimentation often provides the best and most expensive information available about the performance of the design. Because experimentation can be so costly, it should be well-planned. Some suggestions for getting more value out of your experiments are given in the Development Reference under Experimentation (11.26).

Good experimental skills are often a key discriminator for excellent designers.

3.8 Evaluating

Evaluating is the skill of integrating all the information at hand to determine the quality of a design outcome (which may include the entire product). In general, the quality of a design outcome will be determined by its desirability and transferability. Key actions that are part of this skill include comparing, applying judgment, evaluating objectively, and concluding.

Throughout the product development process, the desirability and transferability of the design are evaluated. The best designers frequently evaluate the team's work, including their own. Frequent honest evaluation leads to improved performance.

The specific types of evaluation tend to vary depending on the stage of development.

Evaluation Types

There are three types of evaluation that we often use during the product development process:

1. Intuition-based evaluation

   Characterized by: Judgment made without the need for conscious reasoning

   Representative activity: Multivoting

2. Qualitative evaluation

   Characterized by: Judgment made based on a qualitative evalution of the alternatives, which generally involves subjective assessments that by their nature have limited repeatability.

   Representative activities: Screening and scoring matrices

3. Quantitative evaluation

   Characterized by: Judgment made based on a generally objective measurement of the alternatives. In general, quantitative evaluations have less variability and more repeatability than qualitative evaluations.

   Representative activities: Lab testing, FEA, engineering analysis

It is valuable to understand that these types of evaluation exist, because each of them requires a different level of time and detail to carry out. Intuition-based evaluation generally requires less information and time than qualitative evaluation; and qualitative less than quantitative evaluation. Understanding this helps us plan when and how to use these evaluation types. It is best to plan in advance the types of evaluation that will be used to make specific design decisions.

Table 3.1: Typical evaluation activities used in various development stages.

Opportunity Development	Concept Development	Subsystem Engineering	System Refinement	Producibility Refinement	Post-release Refinement
Cost-benefit analysis Multivoting	Objective Tree Multivoting Scoring and Screening Matrices Feasibility Judgment Axiomatic Design	Finite Element Analysis Computational Fluid Dynamics Design of Experiments Failure Modes and Effects Analysis (FMEA) Axiomatic Design	Fault Tree Analysis (FTA) Sensitivity Analysis	C_{pk}, C_p analysis Process FMEA	Design of experiments FMEA C_{pk}, C_p analysis Process FMEA

Evaluation Activities

A number of different evaluation activities are used to evaluate the different alternatives and support decision making in product development. Some of the typical evaluation activities for each of the stages of development are listed in Table 3.1. Many of these activities are described more fully in the Product Development Reference.

3.9   Deciding

Deciding is the skill of choosing a course of action based on the currently available information. As previously discussed, product development is the process of advancing the design of the product until it is ready to be used for manufacturing the product in the desired quantities and with appropriate quality and performance. Key actions that are part of this skill are choosing and committing.

The fundamental act of advancing the design is making decisions that increase the level of detail captured in the design. Existing as merely an idea, there are yet infinite possibilities of specific products that could be developed to meet the market requirements. As a production-ready product, on the other hand, every detail of the design is specified so that a third party with the appropriate skill can manufacture identical products based on the design.

Even though it is crucial to advancing the design, product developers may be hesitant to make decisions, because they feel like they don't have enough information to make the right decision. However, all the desired information is never available. Avoiding or postponing decisions is avoiding or postponing progress on the project.

This section provides some guidelines to help you make effective decisions in a timely manner.

Mundane and Vital Decisions

Not all decisions are created equal. There are a few decisions (approximately 10-20%) that will have a strong influence on how well the design meets the market needs. These decisions, called the *vital few*, will take 80-90% of the design time. Because these are so crucial, they require an optimal solution, rather than just any solution. Therefore, they should be made with the best possible design and decision-making practices. For example, if you are designing an automatic transmission for a car, the gear ratios in the transmission are likely to be part of the vital few.

Most of the decisions (approximately 80-90%) have relatively minor influence on how well the design meets the customer needs. We call these needs the *mundane many*. Because they have a small influence, they don't need to be optimal; satisfactory is good enough. The goal in making these decisions is to spend as little time as possible to develop an acceptable solution. In transmission design, the bolts holding the case together are likely to be part of the mundane many.

One of the key decisions you will make in the design process is the decision about which choices are vital and which are mundane. The remainder of this section deals with decision making for the vital few.

Motivation for Decision Making

Decision making is fundamentally a human activity. Although recommended tools can help in the decision-making process, ultimately people, not the tools, make the decisions. The motivations of the people involved can affect the process. There are at least three levels of motivation for decision making in product development, as shown in Table 3.2.

Some of these motivations lead to decision making that is good for the product, while others can lead to decisions that neglect the product. In general, decisions that are made with the success of the product explicitly considered are likely to lead to a better product, and are also likely to lead to most of the positive outcomes identified in Table 3.2.

Decision-Making Processes

Decision-making processes can be classified into implicit and explicit methods.

For implicit methods, the process used to arrive at a decision is not clearly specified. It often seems as if the decision just happens. Common pitfalls associated with implicit processes include the following:

- Everybody agrees to the first proposal, without expressing true feelings

- Go with the loudest person's recommendation

- Go with the most senior person's recommendation

- Let the client or manager make the decision

- Follow the most persuasive person's opinion

- Accept the choice of the most tenacious team member

- Choose the idea with the most detail.

Each of these implicit methods is dangerous to use, because there is no clear understanding of how or why a decision was reached. However, good products can, at times, be developed using implicit decision methods.

Explicit decision methods require the team to decide how a decision is to be made. Decisions can be made by individuals, by the team, or by a subteam. For decisions made by individuals, choosing which person will make the decision is important in an explicit process. The decision maker might be the boss, the client, the team member with the most expertise, the team member who feels most strongly about the decision, or a team member who has been given the assignment to make the decision. But in each of these cases, the reason *why* an individual is making the decision has been explicitly declared.

Individual decision making is likely to be a good choice when the decision is to be made between a few very good candidates, or when the choice is part of the mundane many. However, if the choice is part of the vital few and there are some inferior candidates, individual decision making may be less effective than a team-based decision.

When a team chooses to make a decision, there are two fundamental methods: voting and consensus. Voting requires individuals to decide on a preference, and then the candidate decision with the majority of the votes will be selected. Voting can be an efficient way of deciding things, but it can also be dangerous. The dangers of voting include alienating team members on the "losing" side of the vote, and failing to come to a common understanding of the

Table 3.2: Representative motivations for decision making.

Individual Motivations	Team Motivations	Organizational Motivations
Get promoted	Minimize risk of failure	Develop product development capability
Avoid conflicts with team members	Maximize product performance	Strengthen team cooperation
Minimize time spent on decision	Please management	Develop successful product
Avoid responsibility for the decision	Develop a successful product	Improve individual capabilities
Do a good job on the project	Maintain good working relationships	Let somebody else accept the responsibility

solution candidates. However, voting can be a good process for making some of the less-important decisions.

Consensus is likely to be the slowest way to make a decision, but usually results in the best decisions. Consensus requires the individual team members to agree that a given decision is the best. Note that consensus is not compromise. Compromise implies that two people who disagree find an alternative that makes both of them somewhat unhappy. In contrast, consensus requires two people who disagree to find an alternative that both can fully support.

One tool that is often used to help build consensus is the evaluation matrix (sometimes called a decision matrix). Evaluation matrices can be used throughout the design process, but it is common to use evaluation matrices in concept selection, such as the Concept Screening and Concept Scoring matrices. A related, but somewhat different use of evaluation matrices is the Controlled Convergence method advocated by Pugh (1991, pp. 74–85).

Evaluation matrices do not make decisions. They simply present the results of evaluations to help individuals and teams make decisions.

Decide to Decide

One of the major mistakes made by product development teams is to unnecessarily postpone making design decisions, assuming that if they wait just

another day or two they will have all the information they need to decide. As long as decisions are not being made, the design is not moving forward. Be aggressive about making decisions. Decide, test, and validate. This is a key to effective progress in product development.

3.10 Conveying

Conveying is the skill of sharing the design information in a way that meets the needs of others and advances the design, often by facilitating appropriate evaluation by people both inside and outside the team. Key actions that are part of this skill include salesmanship, writing, speaking, presenting, summarizing, and advocating.

The design team is very familiar with the strengths of the design, and readily sees why it is desirable and transferable. Those outside the design team are less familiar with the product, and are generally less enthusiastic about the product. It is the responsibility of the design team to convey their successes (and concerns) to those outside the team who are stakeholders in the development project.

Effective conveying requires the team to understand the needs of the audience and prepare effective communication materials to meet those needs. The team must clearly share the benefits of the product, while being honest about its limitations. The team should be advocates for their work. Remember that rapid and effective progress is dependent on having the necessary resources. The resources are most often made available based on the perception of management concerning the benefits of the project. By effectively advocating for the project, the development team can ensure the availability of the necessary resources to succeed.

An excellent design that is poorly conveyed is at risk of being judged as a poor design. Thus, for your designs to be approved for production, you must develop the skill of conveying the design to those who make decisions about implementation.

3.11   Exercises

Test Your Knowledge

T3-1 List the crucial design skills.

T3-2 State the goal of prototyping.

T3-3 State the goal of experimenting.

T3-4 State the goal of evaluation.

T3-5 List three types of evaluation that are often used during product development.

T3-6 State the goal of deciding.

T3-7 State the goal of conveying.

Apply Your Understanding

A3-1 In which of the crucial design skills are you the strongest? How have you developed your strength in these areas?

A3-2 In which of the crucial design skills are you the weakest? What plans do you have to improve your skills in these areas?

A3-3 Describe the difference between mundane and vital decisions.

A3-4 Explain why explicit decision methods are generally better than implicit decision methods.

A3-5 Explain one strength and one weakness of deciding by voting.

A3-6 Explain the difference between compromise and consensus.

A3-7 Make a sketch of a solution for carrying a notebook computer safely on a bicycle. Show your sketch to three different people to see if they understand it.

A3-8 Make an engineering drawing of a wooden pencil. Ask someone familiar with engineering drawings to check your drawing and give you feedback.

A3-9 Using a 3-D modeling system, make a solid model of an actual wooden pencil.

    a) Use the system to predict the mass of the pencil.

    b) Weigh the pencil and see how well the calculated weight matches the actual weight.

    c) What do you believe is the major source of the discrepancy between the calculated and actual weights? What evidence do you have for your belief?

A3-10 Using a 3-D modeling system, create a rendering of a solid model suitable for presentation. You may use any solid model available, including one downloaded from the internet.

A3-11 Poor decision making: Think of a time when you have been involved with a team that had poor decision-making practices.

    a) What do you believe was the main dysfunction exhibited by the team?

    b) What could you have done to help overcome this dysfunction?

A3-12 Good decision making: Think of a time when you have been involved with a team that had good decision-making practices.

    a) What do you believe was the key practice that led to good decision making?

    b) How might you replicate this practice in future decision-making opportunities?

A3-13 Assume you are just beginning to develop concepts for a device to hold a smartphone on the handlebars of a bicycle to use as a GPS.

    a) List a few possible engineering models that might be helpful in this stage of development.

    b) For one of these models, develop and solve it.

A3-14 Make a low-fidelity prototype of the smartphone holder referenced in Exercise A3-13. What is the purpose for this prototype? How well did it meet the purpose?

A3-15 Thoughtfully create a design portfolio that conveys your skills in a few or all of the design skills discussed in this chapter. Treat the portfolio like a product and use the design skills discussed to create an excellent portfolio.

4

Opportunity Development

The next four chapters of this book focus on the stages of product development. As a detailed guide, they describe the purpose of each stage, the development outcomes you'll produce and have approved, and some useful tools[1] aimed at helping you evolve the design through each stage of development. We also identify some common pitfalls to avoid in each stage.

When we discuss the development outcomes, we focus on the required information for each stage of development. As the design evolves, new kinds of information will be created and will need to be captured in transferable artifacts. The kind of information discussed is consistent across varying industries and companies, but the artifacts can vary. Thus, we focus on the *kind* of information in these chapters, and only indicate typical artifacts as possible ways of conveying the information to others. By understanding and applying the principles of the required information for each stage, you will be able to make good decisions about the artifacts you create.

We devote four chapters to these stages because *all* manufactured engineered products pass through the six distinct stages discussed. As such, you *will* work in one or more of these stages during product development. So it's worth spending time to understand them.

This chapter discusses opportunity development, which is the first stage of product development. Generally speaking, during the opportunity development stage the team works to evolve information that characterizes the problem being solved. The purpose of this stage is to combine this and other information to define the opportunity.

4.1 Design Evolution During Opportunity Development

The initial idea, which is often sparked when considering the needs and circumstances of people and organizations, marks the beginning of this stage. Key characteristics of the project to develop the idea are captured in the *project objective statement*[2], which is a brief summary of the scope, schedule, and resources of the product development project. The project objective statement serves as a foundation supporting all of the work on the project.

---

[1] These tools are listed in the chapters and described in more detail in the Product Development Reference.

[2] Sometimes called the design brief or mission statement. See Project Objective Statement (11.49) in the Development Reference for more information.

© Springer Nature Switzerland AG 2020
C. A. Mattson, C. D. Sorensen, *Product Development*,
https://doi.org/10.1007/978-3-030-14899-7_4

Table 4.1: Summary of the opportunity development stage.

Opportunity Development: Develop clear statements of market and engineering requirements that capture the market's desires for the product.

	Required information	Typical artifacts	Checking criteria	Approval criteria
Requirements	Market requirements	Section A of the requirements matrix	Consistent level of generality? Capture the most important requirements? Appropriate number of requirements? Reasonable differences in importance?	Complete and appropriate as evaluated by market representative
Requirements	Performance measures	Section B of the requirements matrix	Clearly measurable (even for subjective)? Capture market requirements well? Generally dependent, rather than independent? Units given and appropriate? Number appropriate? Appropriate importance ratings?	Market representatives (for less technical measures) and/or project approvers (for highly technical measures) find the measures to be appropriate.
Requirements	Requirement-measure correlations	Section C of the requirements matrix	All requirements have at least one measure? All measures have at least one requirement? More than just one-to-one correlations? Appropriate, defensible correlations?	Judged appropriate by project approvers
Requirements	Ideal values	Section D of the requirements matrix	Values make sense? Values are consistent with market requirements?	Judged appropriate by project approvers and market representatives
Tests	None required	Reports of team interactions with market representatives (e.g., surveys, interviews, etc.)	Are the interactions accurately conveyed in the report?	Not approved directly, but used to support approval of the requirements
Models	None required	Simple models that relate desirability to measured performance of competitors (e.g., screen size, battery life)	Do the models make logical sense?	Not approved directly, but used to support approval of the requirements
Prototypes	None required	Rough prototypes (foamboard, paper, foam, clay, cardboard, plywood, etc.) used to communicate with market representatives. Don't fully reflect eventual product design.	Do the prototypes facilitate communication with market representatives?	Not approved directly, but used to support approval of the requirements
Design	None required	Rough sketches or drawings of competitors or generic product possibilities. Don't fully reflect eventual product design.	Do the sketches or drawings facilitate communication with market representatives?	Not approved directly, but used to support approval of the requirements
	Useful tools:	Basic design process, competitive benchmarking, financial analysis, focus groups, interviews, observational studies, patent searches, planning canvas, project objective statement, quality function deployment, requirements matrix, surveys.		
	Common pitfalls:	Assuming, not validating; using only subjective performance measures; delaying feedback; devaluing the opportunity development stage; spending too much time.		

Progress is made during the opportunity development stage as the team gathers information about the market needs and conditions, and comes to understand and articulates what would be required of a desirable product.

To be clear, a well-defined opportunity comprises the following information:

1. Market requirements, which define the requirements for a desirable product in terms the market understands and uses.

2. Performance measures, which are characteristics of the product that will be measured by the development team.

3. Requirement-measure correlations, which identify the performance measures that have a significant effect on each market requirement.

4. Ideal values, which define the values that the market would prefer for the performance measures in the absence of trade-offs.

While the items listed above can be documented in a variety of ways, we find it extremely convenient to capture and present them in the form of a *requirements matrix*[3], which is described in detail in the Product Development Reference. Therefore, we will use requirements matrices throughout this book to discuss and present the opportunity definition.

As a guide to the opportunity development stage, we have prepared Table 4.1. This table summarizes the purpose, development outcomes, approvals, useful tools, and common pitfalls that characterize the opportunity development stage. These items are explained more fully below.

---

[3]An example of a requirements matrix is shown in Figure 4.2.

Purpose

The purpose of the opportunity development stage is to create clear statements of market and engineering requirements that capture the market's desires for the product.

At the end of this stage, these elements are checked and validated to ensure they are thoughtfully articulated, unambiguous, and accurately reflect the market's expectations.

Development Outcomes

The following development outcomes will be achieved as the design evolves through the opportunity development stage. During opportunity development, as in each development stage, development outcomes may be part of the requirements, tests, and design; or they may be models and prototypes that are used to predict and/or demonstrate the desirability of the design.

Requirements

There are four essential elements of the requirements during opportunity development.

1. Market Requirements

   Market requirements clarify what is needed for the product to be desirable to the market. They are objective and subjective product characteristics that customers use to make decisions about the products they will purchase. Thus, market requirements identify key factors that determine desirability.

   For the maximum benefit, there will typically be from five to twenty requirements. If there are too few requirements, either some are left out or the requirements are too general to be of much use in making design decisions. If there are too many requirements, it is difficult to consider them all during the development process. And in fact, consumers seldom use more than a few criteria to decide on their purchases.

   All of the requirements should be expressed at approximately the same level of generality. They should also be expressed in the form of product-focused requirement statements. To develop this succinct but representative list of market requirements, follow the guidelines laid out in the Development Reference under Requirements Hierarchy (11.54) and Product-Focused Requirement Statements (11.48).

It may be desirable to list a relative importance for each of the market requirements. This relative importance helps the team focus on the most important requirements. The relative importance is sometimes indicated as a percentage (especially when there is a small number of requirements). It can also be given a geometric scaling such as 1 (3^0) for low importance, 3 (3^1) for moderate importance, and 9 (3^2) for high importance.

Assigning importance to requirements is necessarily a subjective process. After importance has been assigned, the team should carefully check whether the listed importances actually reflect the opinion of the market.

Because the requirements capture the desires of the *market*, rather than the desires of the team, market representatives must be involved in the approval process for the market requirements.

2. Performance measures

By definition, the market requirements reflect the concerns of the market. Also by definition, the development team is not the market. Therefore, in order for the team to develop a desirable product, they must have some means of determining whether the product is desirable. The performance measures are product characteristics that can be measured by the team and provide information about the desirability of the product.

Where possible, performance measures should be quantitative. However, in some cases it is necessary to use qualitative performance measures. In these cases, a

specific description of the desired qualitative performance can help minimize subjectivity of the evaluation. Rubrics for performing the qualitative evaluation can provide a quantitative assessment for qualitative performance.

There should be a unit of measurement associated with each performance measure to help clarify what is being measured and to avoid confusion among team members and other stakeholders.

Performance measures should generally be dependent, rather than independent, characteristics of a product. This is simply because if the designer is free to independently choose some characteristic, it's not really a performance measure, and the design team can just choose a desirable value. Instead, performance measures should require that the designer choose proper values of some other independent parameters in order to assure that the performance value is met.

In order to help the development team focus on the most important measures, it is often useful to have an importance rating for each of the performance measures. The importance rating may be assigned by the team in consultation with market representatives. It may also be obtained by combining the importance of the market requirements with the requirement measure relationships (see Quality Function Deployment (11.51) in the Development Reference for more on how this is done).

Some performance measures are easily understood by market representatives (e.g., fuel economy). Such measures should be approved by market representatives before the end of opportunity development. Other highly technical measures need only be approved by the project approvers.

3. Correlations between market requirements and performance measures

Performance measures are often developed by considering ways to measure how well the product meets a particular market requirement. In such cases, there is a strong correlation between the market requirement and the performance measure. However, it may take several performance measures to adequately measure the achievement of some market requirements. Similarly, a single performance measure may help to determine multiple market requirements.

The correlations between market requirements and performance measures can be easily captured in matrix form, such as that shown in Figure 4.2, where a dot indicates the performance measure in the column is correlated with the market requirement in the row.

The correlations captured here should be *measurement* correlations. For example, consider the design of a car, with a market requirement of good gas mileage and performance measures including vehicle weight and EPA city fuel economy rating. EPA city fuel economy rating is a measure of how good a vehicle's gas mileage is, so these have a measurement correlation. Measurement correlations are captured in Section C of the requirements matrix.

In contrast, vehicle weight is not used to measure gas mileage, even though weight is correlated with gas mileage. Because this is not a measurement correlation, it is not captured in Section C of the requirements matrix.

All market requirements should have at least one performance measure, or else they will be largely unassessed by the team.

Performance measures without market requirements indicate either an incomplete understanding of the market requirements or an irrelevant performance measure.

Generally some measures are correlated with multiple requirements. Also, many market requirements require more than one performance measure. A strict one-to-one correlation between market requirements and performance measures is likely to indicate a superficial job of identifying correlations.

4. Ideal values for performance measures

These values represent the desires of the market for the performance measures in the absence of trade-offs. For each performance measure there should be an ideal value, which indicates the value the market most prefers. There may also be upper and/or lower acceptable limits, which indicate the limits to the values that the market will find acceptable. Values outside of the acceptable limits will lead to an undesirable product. It is possible that either the upper or lower acceptable limit does not exist. In such cases, this should be explicitly indicated.

Tests

No tests (of the design) are required during opportunity development, simply because the design has not yet started to take form in this stage of development. However, as performance measures are created, it is useful to consider tests that can be used to evaluate the performance of a design that will begin to emerge in the next stage of development.

Interaction with market representatives (surveys, interviews, etc.) can be considered tests of the market requirements as they are developed. Such interactions generally lead to better market requirements.

Models

No models (of the design) are required at this stage of development. However, market understanding is often facilitated by the creation of simple technical models that relate desirability (e.g., sales, product ratings) to measured performance of competitive or existing products (e.g., screen size, battery life).

Prototypes

No prototypes (of the design) are required at this stage of development. However, mock-ups (foamboard, paper, foam, clay, cardboard, plywood, found items, etc.) can often be used to improve communication with market representatives and obtain better market requirements information.

Design

No design — meaning no definition of the product — is required during opportunity development. However, it is often desirable to create sketches of generic product possibilities to use in interactions with market representatives. Because these artifacts are not focused on the intended product, but rather on possibilities, they are not part of the design (i.e., they don't define the product being developed).

Useful Tools

To advance the design through this stage of development, the product development team completes design activities. To help you get started, and for later reference, we list the following tools as often useful during the opportunity development stage:

For information gathering: Benchmarking (11.2), focus groups (11.31), interviews (11.34), observational studies (11.40), patent searches (11.43), surveys (11.65), literature search.

For establishing the market requirements: Quality function deployment (11.51), requirements matrix (11.55), requirements hierarchy (11.54), product-focused requirement statements (11.48), goal pyramid (11.32).

For general design work: Basic design process (11.1), financial analysis (11.29), planning canvas (11.47), project objective statements (11.49), storyboards (11.64).

Common Pitfalls

Having interacted with or participated on numerous product development teams, we're familiar with the pitfalls of this stage of development. Here's a short list of common pitfalls to avoid:

- Teams assuming — not validating — what the market wants: Avoid substituting team, supervisor, sponsor, or end-user judgment for all stakeholder judgment. Avoid the risk

of assuming what the market wants. Seek and listen to honest feedback. Unproven assumptions cause significant problems.

- Teams using only subjective performance measures: Avoid converging on only subjective performance measures. Work to define these items to the point that most can be measured with objective performance measures.

- Teams trying to refine and perfect their work before seeking feedback: Embrace the concept of failing often to succeed sooner (Kelley and Littman, 2001, Chapter 15).

- Teams failing to see the value of this early stage of design: The trajectory for the product and the project is established in this stage of development. Spending too little time or giving too little attention to this stage of development means having a higher risk that the final product is not wanted by the market.

- Teams spending too much time on the design activities: More time can always be spent on a stage of development; avoid endless development by passing through this stage of development quickly, but with a sufficient understanding of the market requirements to do a good job at subsequent stages of development.

4.2 Example of Opportunity Development

To clarify the product development artifacts that result from the opportunity development stage, an example product is presented. The product chosen is a human-powered water well drill for the developing world, which is shown in finished form in Figure 4.1, and is described in detail as a case study in Appendix B (found on page 327). This example will be revisited for each of the stages of development, so you may find it valuable to review Appendix B now.

The approved project objective statement for the human-powered drill is:

Design, build, and test a human-powered drill that reaches underground potable water at depths of 250 ft in all soil types by March 25, 2011 with a prototyping budget of US$2,800 and for less than 1,700 man hours of development.

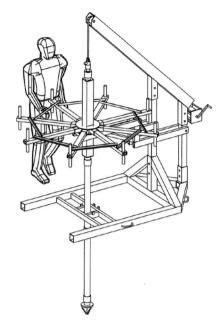

As listed in Table 4.1, the requirements at the end of the stage must include a defined opportunity, which can be captured in Sections A, B, C, and D of the requirements matrix. Figure 4.2 shows these sections of the matrix for the drill at the end of opportunity development[4]. As seen in the figure, the market requirements are listed on the left-hand side. This list represents *what* the market requires. For this example, the team identified the list by speaking with the client and experts, by benchmarking, and by using other forms of research.

After establishing the market requirements, the team considered them one-by-one, and chose performance measures for each requirement. These measures define the attributes of the product that will be measured or evaluated to determine whether the design meets the market requirement. For example, the market requires that the drill reach potable water beyond 100 ft (market requirement #1). Considering this requirement alone, the team chose how it might be measured or evaluated. The team chose maximum borehole depth (in units of feet) and time required to cut through 6 inches of rock (in units of minutes) as the relevant performance measures. The relationship between performance measures and requirements is captured by the dots placed in the relationships section of the matrix.

By choosing these performance measures, the team indicates that the market requirement for reaching potable water beyond 100 feet will be met if the drill

Figure 4.1: Model of the finished human-powered water well drill. This drill resulted from an engineering effort to bring clean drinking water to rural communities in sub-Saharan Africa. The device was designed to drill a 6 inch diameter hole 250 feet into the ground to access clean drinking water.

achieves desirable values of maximum borehole depth and time to cut through six inches of rock.

The team carried out this process for each of the market requirements. To minimize the risk that they are not representative of the market requirements, all product development team members should ask themselves if they believe the performance measures to be correct — that this set of measures adequately represents the market requirements.

> *Check the requirements: Do you think the performance measures adequately represent the market requirements for the human-powered water well drill? Would you have chosen additional or different performance measures?*

---

[4]As described in the Product Development Reference, the full requirements matrix includes the real values at the bottom of the matrix, and the market response on the right side of the matrix. These two elements are not developed as part of this stage.

Product: DRILL
Subsystem: N/A

Performance Measures (and specifications)

#	Performance Measure	Units	Importance	Upper Acceptable	Ideal	Lower Acceptable
1	Maximum borehole depth	ft	9	–	250	100
2	Time required to cut through 6 inches of rock	min	10	60	45	–
3	Downward drilling force	lbs	10	–	3,000	500
4	Torque applied to drill bit	ft-lbs	10	–	400	200
5	Compatable with X% existing drill bits	%	3	–	100	90
6	Water pressure down the pipe	psi	6	–	113	50
7	Percentage of water that leaks through sides	%	3	5	0	–
8	Percentage volume of cuttings removed	%	3	–	100	95
9	Depth cut per 8 hours of drilling	ft	3	–	36	4
10	Number of required people	people	9	12	3	–
11	Weight of heaviest subassembly	lbs	9	400	50	–
12	Longest dimension (l,w,h) of biggest subassembly	in	9	96	48	–
13	Percentage of drill manufacturable in Tanzania	%	9	–	100	85
14	Cost to produce 1 drill after development	USD	9	5,000	1,000	–
15	Time required to learn how to operate	hr	9	20	4	–
16	Height of hand operated parts	ft	1	5	3.5	2
17	Feels comfortable	n/a	1	–	Drill can be operated continuously without the need to rest. Does not require awkward movements.	Drill can be operated with occasional rest, and requires awkward movements that leave the user sore.
18	The Drill is attractive	n/a	4	–	Drill has a professional look. People are interested in looking at it.	Drill looks like a piece of machinery for drilling holes.
19	The Drill interests investors	n/a	12	–	Drill captures media attention. Investor s are proactive in contributing. Drill has iconic look.	The drill, when explained to investors, is something they want to invest in.

Market Requirements (Whats) — Relationship Matrix (● = relationship)

#	Market Requirement (Whats)	Importance	1	2	3	4	5	6	7	8	9	10	11	12	13	14	15	16	17	18	19
1	The Drill reaches potable water beyond 100 ft	9	●	●																	
2	The Drill cuts through rock	1		●	●	●															
3	The Drill uses existing drill bits	3					●														
4	The Drill seals borehole sides to prevent cave-in	3						●	●												
5	The Drill removes cuttings from the borehole	3						●		●											
6	The Drill works at an efficient speed	3									●										
7	The Drill uses only manual labor to function	9			●	●					●										
8	The Drill is affordable	9											●	●							●
9	The Drill requires simple training to operate	9															●				
10	The Drill is portable	9											●	●							
11	The Drill is comfortable to operate	1																●	●		
12	The Drill is attractive	1																		●	
13	The Drill attracts investors	3																		●	

Figure 4.2: The system requirements matrix at the end of the opportunity development stage for the human-powered water well drill.

Once the performance measures are established, they become excellent talking points for deeper conversations with the client, experts, and end users. They are also specific enough that competitive products can be benchmarked against them. These talking points and competitive benchmarking results become a useful part of choosing ideal values and acceptable limits for each performance measure. The acceptable limits and ideal values establish bounds within which the design should be created.

4.3 Top-Level Activity Map for Opportunity Development

The previous sections have shown the outcomes of the opportunity development stage, but have said little about the activities and sub-outcomes used to achieve these outcomes. To that end, this section provides a relatively high-level activity map that can be used to produce the main outcomes for opportunity development.

While we provide this section as a guide, it is important to remember that it is the product development team's privilege and responsibility to choose the specific arrangement of design outcomes and the design activities that will lead to them. Ultimately, the specific outcomes and actions chosen need to meet the unique needs and conditions of their project, client, and/or product development team.

Figure 4.3 provides a top-level activity map for the opportunity development stage. This activity map can be thought of as a somewhat more detailed look at what goes on during the opportunity development stage.

Notice that we have not specified definitive activities that should take place. We have, however, provided some example activities that could be used to accomplish the outcome. The fact is that there are multiple design activities that could be used to achieve the outcomes. Many of the example activities listed here are described in the Product Development Reference.

Compound Outcome 5 in Figure 4.3 is an approved project objective statement. Crafting and revising this statement helps the team understand the scope of the project. Notice that this outcome is achieved by completing multiple activities including *Interacting with the client*, *Creating and revising a project objective statement*, and *seeking approval*.

Recognize that the project objective statement is achieved through significant interaction with the client, while Outcome 6 (market/customer statements) is focused on interactions with people who represent the market — which are typically not the client. Also notice that Outcomes 9 (market requirements) and 10 (performance measures) are not fundamentally about interactions with clients nor market representatives, but are worked through within the team as the market statements are processed. Although the initial market requirements and performance measures are independent, the final requirements (Outcome 15) and performance measures (Outcome 14) are interdependent as indicated by the double-headed dashed arrow. This means that market requirements and performance measures affect each other.

To solidify this concept, consider market requirement #6 for the drill, and performance measure #9 as shown in Figure 4.2. Requirement 6 is *the drill works at an efficient speed*. The ideal value for performance measure #9 is that the drill cuts 4.5 feet per hour. Now, which was identified first: work at efficient speed, or cut 4.5 feet per hour? If the team chooses to serialize, either one could have come first. When the market requirement comes first, the team chooses one or more performance measures to represent it. When the performance measure comes first, the team must discover what (often unspoken) market requirement(s) it represents.

To handle the activities that lead to the interdependent outcomes in Figure 4.3 (Outcomes 13, 14, and 15), the team first develops an initial requirements matrix

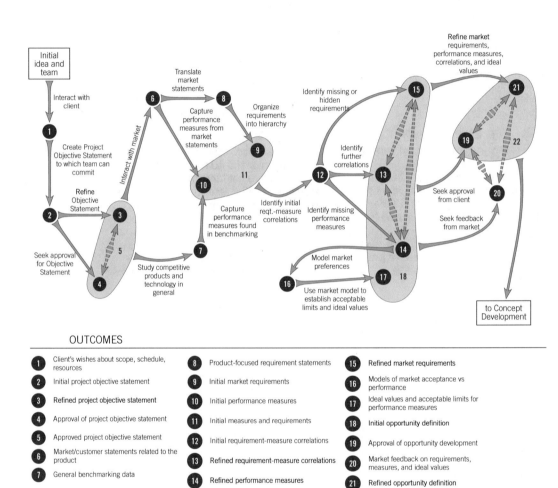

OUTCOMES

1	Client's wishes about scope, schedule, resources	**8**	Product-focused requirement statements	**15**	Refined market requirements
2	Initial project objective statement	**9**	Initial market requirements	**16**	Models of market acceptance vs performance
3	**Refined project objective statement**	**10**	Initial performance measures	**17**	Ideal values and acceptable limits for performance measures
4	Approval of project objective statement	**11**	Initial measures and requirements	**18**	**Initial opportunity definition**
5	Approved project objective statement	**12**	Initial requirement-measure correlations	**19**	Approval of opportunity development
6	Market/customer statements related to the product	**13**	**Refined requirement-measure correlations**	**20**	Market feedback on requirements, measures, and ideal values
7	General benchmarking data	**14**	**Refined performance measures**	**21**	Refined opportunity definition
				22	Approved opportunity definition

Figure 4.3: Top-level activity map for the opportunity development stage.

using information obtained from benchmarking and customer statements. The team then may do either of the following:

- Complete the activities "Identify missing or hidden requirements," "Identify further correlations," and "Identify missing performance measures" in any order, one at a time, repeating until no further items are identified.

- Complete the activities listed above in parallel, simultaneously progressing on all three until no further items are identified.

If the team chooses to run the activities in parallel, it is wise to choose a short coordination interval. This allows all three activities to be executed to result in preliminary outcomes, which are then examined together. All three activities would then be re-executed, if needed, to result in improved outcomes. This process would continue until all three outcomes (13, 14, and 15) are desirable. In this way the activities are carefully coordinated.

We can also see in Figure 4.3 that Outcome 9 (market requirements) and Outcome 10 (performance measures) are compound, which is shown as a gray oval comprising both outcomes and labeled Outcome 11. This graphical representation means that the activity "Identify initial requirement-measure correlations" requires both the initial market requirements and the initial performance measures as input. Notice that this is subtly different from "Model market preferences," which only needs Outcome 14 as input (represented by the arrow coming from Outcome 14 and not from the compound node), even though Outcome 14 cannot be complete until both Outcomes 13 and 15 are complete.

Continuing in the map, we see that Outcomes 13-15 and 17 are all needed before "Seek approval from client" can begin, which makes them a compound outcome (labeled Outcome 18). In this

specific context, the team seeks approval for the market requirements, the performance measures, the correlations between them, and the ideal values. Together, these define the opportunity.

Ultimately this activity map for the opportunity development stage results in an approved opportunity (Outcome 22), which is exactly what the team will need when beginning the concept development stage.

Remember that Figure 4.3 is just one possible activity map; others could be developed to accomplish the same high-level outcome. We reiterate that within this activity map, there is substantial freedom for the team to customize the process by further decomposing and choosing more specific design activities.

Recognize that the details in the activity map are not superfluous, but are the details that will help you plan and carry out your product development work.

4.4   How to Gather and Process Essential Information

One of the most important things to recognize about the opportunity development stage is that it is not merely about formalizing the opportunity, it's about *developing* it. What this means is that our understanding — and the client's understanding — of the market requirements, performance measures, requirement measure relationships, and the ideal values will grow as the opportunity evolves.

The activity map provided in Figure 4.3 shows the sub-outcomes of the stage; but simply getting something for those outcomes will not be enough. We'll want to develop the opportunity as effectively and efficiently as possible. To do that, we will need to be thoughtful about how we choose to gather and process information. To this end, we use this section to describe a few things we like to keep in mind when working through the opportunity development stage.

Representing the Market

It is impossible to know exactly how the market will respond to a new product, without actually putting it on the market. However, it is possible to obtain information from potential customers, users of competing products, experts in related fields, marketing experts, etc. These sources of market information are collectively referred to as market representatives.

For successful product development, it is vital that the market representatives actually reflect the real market. Care in choosing market representatives will be rewarded with higher-quality information about the real market.

When choosing market representatives, it is important to consider the specific market segments at which the product is aimed, rather than the overall market of related products. For example, if a team were designing a new automobile, different market representatives would be used for an economy car and a luxury sports sedan.

Although the development team members should do their best to understand the market, it is very risky to completely substitute the team's judgment for that of the market. The market representatives that are chosen during the opportunity development stage will be used to validate the design at the end of each stage of development. Different individuals may be used, but the collection of individuals should represent the same market population.

Understanding the Market

If we are to have products that the market finds desirable, we must understand the market thoroughly. In this section, we describe some design activities that are often pursued to understand the desires of the market.

The objective in understanding the market is to find out what the market cares about in a product or a potential product. Different techniques are used, with varying breadth, cost, and quality of information. Five techniques that are particularly pertinent here are described briefly below, with more details found in the Product

Development Reference. They are: surveys, focus groups, interviews, observational studies, and product benchmarking. In general, product development teams will use all or most of these techniques in a strategic mix as they seek to understand what the market wants.

Surveys

Surveys obtain written feedback about customer preferences. They may be performed in various ways, such as by telephone, post, email, website, or in person. Surveys generally are the cheapest way to obtain information and are therefore used to sample a very broad section of the market. A major limitation of surveys is the inability to ask effective follow-up questions to better understand the responses that are given.

Focus Groups

Focus groups are group interviews of small groups of potential customers. In a focus group, a facilitator will ask the participants to discuss the issues relating to the product or the needs leading to the product. Focus groups will often use props such as competitive products to help the conversation become more real and focused. Because the facilitator is interacting with the group members, follow-up questions can be used to better clarify and understand any concerns or needs expressed.

Focus groups are limited to relatively small groups, so multiple focus groups will be held if a large market sample is desired. Information obtained from focus groups is generally of higher quality than survey information, but comes from a less broad spectrum of the prospective market.

Interviews

Interviews are one-on-one discussions with potential customers. Props such as competitive products can be used to help spur and focus the conversation. In interviews, the subjects are not influenced by others (as they are in focus groups), so information may better reflect an individual's feelings. On the other hand, the lack of others in the interview may result in a subject overlooking some important issues.

To capture effective market information, the number of interviews required is significantly larger than the number of focus groups. The cost of an interview is only marginally less than the cost of a focus group. For this reason, focus groups are often preferred to interviews.

However, quiet individuals are more free to speak in interviews than in focus groups. Interviews are also free from groupthink and from influence exerted by dominant individuals. Because of these strengths, interviews should generally be part of the process to determine market desires.

Observational Studies

In observational studies, a member of the development team goes to the location of a potential customer and observes what the subject does, rather than what the subject says. A perceptive observer will learn much more about the subject's needs and desires than will ever be expressed by the subject in words. This information tends to be some of the highest quality information that can be obtained.

However, this information is also expensive to obtain. It is likely that subjects will only be interacting with the potential product for a short fraction of the time they are together with the observer. Therefore, the time required to do a single observational study will generally be much larger than the time for an interview.

The quality of observational studies makes them essential; their cost means that they must be used wisely.

Product Benchmarking

To create excellent solutions, it is vital to understand the range of available products. Even for products that are truly groundbreaking, there are alternative ways that the requirements are currently being met. Understanding everything possible about competitive products – called

benchmarking – helps in a variety of ways. Benchmarking can help the development team understand the performance, cost, and technology of competitive products. It can help determine the target performance values and financial implications of the product under development. It can provide valuable seeds for creativity in the development team.

Techniques for discovering and understanding the competitive landscape are described in this section.

Products for competitive benchmarking should be chosen from the leading products of the market segment(s) at which your product is aimed. Determining these benchmark products requires discovering the products that are currently in use. The benchmarking will be no better than the range of other products considered, so it's important to be thorough and thoughtful in choosing the benchmark products.

There are two basic kinds of benchmarking:

- Technical benchmarking: One method of analyzing competitive products is technical benchmarking. In technical benchmarking, the performance measures[5] identified in the opportunity are measured for each of the competitive products. At the end of this activity, the team will understand in detail the performance of the competitive products in the market segment.

- Market benchmarking: A second method of analyzing competitive products is market benchmarking. In market benchmarking, the market's rating of the strengths of the competitive products is obtained. Competitive evaluation can be obtained directly from market representatives, using the techniques listed above such as surveys, interviews, and focus groups. Market ratings can also be obtained from various product reviews, such as magazine reviews and online reviews.

---

[5]One of the four things that defines the opportunity.

In addition, market performance information is often available from published market share analyses. The goal of market performance benchmarking is to thoroughly understand what the market thinks about the benchmark product, both in overall evaluation and in evaluation of the specific market requirements you have chosen.

Taken as a whole, the results of technical benchmarking and market benchmarking provide a valuable source of understanding the desires of the market.

Processing the Information Gathered During This Stage

In order for the gathered market information to be valuable to the team, it needs to be processed and condensed so that it can be continually used by the team to evaluate the desirability of the design throughout the remaining stages of product development. Here, we briefly discuss the processing of information as it pertains to the sub-outcomes of this stage.

Capturing Customer Statements

Customer statements are statements made by individuals who are part of the market representatives about their desires for the product being developed. They should be recorded in the customer's own words. They can be obtained from interviews, focus groups, online surveys, paper surveys, or any other method of hearing what the customer thinks about the product. Capturing the information in a form as close as possible to the customer's actual words is important in order to avoid substituting the development team judgment for the market judgment.

Customer statements tend to be narrowly focused on specific desires, although top-level desires are also expressed at times. The different levels of specificity for customer statements will be rationalized using a requirements hierarchy.

Developing a Requirements Hierarchy

A requirements hierarchy results from translating the customer statements into product-focused requirement statements, then organizing the resulting requirement statements into related groups. For each group, a summary requirement statement is created that represents the more specific requirements in the group.

Table 4.2 lists six guidelines developed from those presented by Ulrich and Eppinger (2012, pp. 81–83) for writing effective product-focused requirement statements. While these guidelines may seem arbitrary, in our experience we think more clearly about the product when we follow the guidelines.

If the number of summary requirements statements is appropriate (5 to 20), the summary requirements become the market requirements. If there are too many summary requirements statements, the organizing process may need to be repeated until a manageable number is reached.

Developing Performance Measures

For each of the market requirements, the team should ask itself "How might we measure this requirement?" Each of the answers to this question will become a performance measure. Beware of the temptation to only get a single performance measure for a given requirement. Because the requirements are summaries of multiple needs, it's likely that there will be multiple performance measures for a given need.

It's also helpful for the team to ask itself "What else might be useful to measure on a competitor's product?" This may lead to additional performance measures that relate to multiple market requirements or that relate to market requirements not yet identified.

Determining Requirement-Measure Correlations

Once the market requirements and performance measures are identified, the

Table 4.2: Six guidelines for writing effective product-focused requirement statements from customer statements. Examples are given for customer statements found in online reviews of cordless circular saws.

Guideline	Customer Statement	Good Requirement Statement	Inferior Requirement Statement
Express the requirement with positive, not negative, phrasing	It just doesn't have the torque of a corded saw and will choke in a slight bind while cutting plywood sheets.	The saw resists binding when cutting plywood.	The saw doesn't bind when cutting plywood.
Express the requirement as specifically as the user statement	I wish it had a case to protect it from dents and dings.	The saw is resistant to dents and dings.	The saw is rugged.
Express the requirement as a requirement of the product, not the environment or the user	The trigger safety interlock is aggravating! I have incredibly large hands and still find the trigger interlock a stretch to reach.	The saw can be started easily by most people, even with safety interlock features active.	People can easily reach the safety interlock.
Express a requirement, rather than a performance measure	The battery lasts for three hours of typical carpentry work.	The battery lasts a long time. *or* The battery life allows typical carpentry work.	The battery lasts for three hours.
Express a requirement, rather than a product feature	Having the blade on the left side gives a right-handed user a great line of sight.	The saw provides clear sight to the cut.	The saw has the blade on the left side.
Express the requirement independent from its importance	The saw does not have a light which I miss.	The saw illuminates the cut area.	The saw should illuminate the cut area. *or* The saw must illuminate the cut area.

team should review the correlations between the requirements and the measures. This can often reveal additional correlations that help the team focus on the most critical measures. The Development Reference entry on Goal Pyramid (11.32) teaches a process for identifying the measures that will have the greatest influence on the success of the product.

A matrix that shows only one-to-one correlations between requirements and measures likely reflects superficial thinking about requirements, performance measures, or both. Also, a matrix that shows correlations between most requirements and measures is probably giving too much credit to weak correlations.

At this point, there may be some market requirements that are not adequately captured by the performance measures. For these requirements, it is important to develop subjective performance measures to ensure that the market requirements are fully accounted for in later decisions.

After determining the requirement-measure correlations, the team should be satisfied that the market requirements are fully captured in the performance measures. To the extent possible, the performance measures should be objective. In addition, subjective performance measures should be used where necessary to fully capture the market requirements.

Determining Ideal Values

The most significant competitive products should be benchmarked according to the performance measures (technical benchmarking). This will provide a list of the performance of competing products. The performance data can then be coupled with the market rating and market share data obtained in market benchmarking to determine how the market responds to various levels of performance. Thoughtful evaluation of this data allows the determination of ideal values and acceptable ranges for each of the objective performance measures.

For the subjective performance measures, a description of each of the competitive products can be written. When compared with the market rating and market share data for the product, a description of ideal performance and the acceptable limits in the subjective criterion can be written.

4.5   Summary

During the opportunity development stage the team evolves the design by capturing market requirements, determining performance measures, defining requirement-measure correlations, and establishing ideal values of performance. Generally speaking, the opportunity development stage is complete when these items are approved.

It is worth noting, however, that companies often require additional *project-related* items[6] to be approved at the end of this and other stages. These can include the approval of a project plan for the upcoming stage of development, a development budget assessment/plan, or the approval of a change management procedure.

Performance measures with their ideal values and acceptable limits are essential to the development process. These will allow the team to estimate the desirability of the design throughout the product development process.

A top-level activity map – comprising 22 outcomes – has been presented as a guide for accomplishing the goals of this stage.

___
[6]As opposed to the *design-related* items presented in this chapter.

Multiple activities were introduced in this chapter that can help the team understand and process the desires of the market.

Thoughtful completion of the opportunity development stage, including the validation of the defined opportunity with carefully selected market representatives, can greatly increase the chances of developing a desirable design.

One of the most important messages of this chapter is that by itself the team does not have enough information to successfully define the opportunity. To succeed, the team must interact with at least the client and other market representatives. It is likely that the team will also need to engage with subject matter experts.

4.6   Exercises

Test Your Knowledge

T4-1 List three characteristics of the project that are included in the project objective statement.

T4-2 List four pieces of information that comprise the *opportunity* as it's developed during this stage.

T4-3 List the necessary requirement, test, and design content at the end of opportunity development.

T4-4 List six guidelines for writing effective product-focused requirement statements.

T4-5 List five techniques that are used to understand the market desires.

T4-6 List two kinds of benchmarking that are used to understand market desires.

Apply Your Understanding

A4-1 Think of an idea for simple consumer product. Write a project objective statement that could be used for a project to develop your idea into a product.

A4-2 In your own words, define a market representative.

A4-3 Come up with an idea for a simple household product. Once you have the product idea, discover six market desires for the product. Describe the desires and the process used to obtain the desires.

A4-4 Imagine you are on a development team tasked with designing a new pen. Discover and understand the competitive landscape for pens.

   a) What classes of pens do you find in the market?

   b) What class of pen would you choose to design?

   c) What are the major competitors in that class?

   d) What are the strengths of the major competitors?

A4-5 One source for customer information is customer reviews on online shopping sites. For this exercise, pick a product with which you are familiar and in which you have interest. Go to an online marketing site (such as amazon.com) and obtain a selection of 5-10 reviews for the product.

   a) Obtain a list of customer statements from the reviews. These statements can be likes, dislikes, or suggestions for improvement. Capture the customer statements in the reviewer's own words.

   b) Write product-focused requirement statements for each of the customer statements you have obtained.

   c) Organize your requirement statements, creating a set of market requirements.

   d) List any new market requirements of which you are aware that did not come from the customer statements. Why do you think these requirements were not included in the customer statements?

A4-6 Take two of the market requirements from Exercise A4-5 and develop an appropriate list of performance measures for these requirements.

A4-7 Validate your understanding of the market requirements from Exercise A4-5 by sharing them with a market representative and obtaining feedback.

A4-8 Write a brief stage report for Opportunity Development that can be used during the stage approval process for a project you're working on. Structure your report so that it answers these two fundamental questions: What are the product requirements for your project (i.e., requirements matrix parts A–D)? And how have you validated that they accurately reflect the market's desires? As a way of supporting the report's claims attach and refer to product development artifacts that the team has produced.

5

Concept Development

5.1 Design Evolution During Concept Development

The second stage of product development is concept development. During this stage, the concept for meeting the design opportunity evolves from a very general idea to a high-quality system architecture, which includes the following items:

1. An accurate, unambiguous, transferable definition of the selected product concept, where enough detail is provided about the spatial and structural relationships of the principal subsystems that basic cost, size, weight, and feasibility estimates can be made. There should also be a transferable demonstration of how the concept's parts and its whole meet the opportunity.

2. The decomposition of the system into major components or subsystems.

3. The definition of the interfaces between the subsystems.

4. Opportunity definitions[1] for each of the subsystems.

5. Target values for the system and subsystems.

---

[1]Sections A, B, C, and D of a requirements matrix.

Generally speaking, during the opportunity development stage the team worked to evolve information that characterizes the *problem* being solved. In the concept development stage the team works to evolve the information that characterizes the *solution* to the problem being solved.

By the end of this stage, the design is checked for desirability and transferability, the predicted performance of the product is compared against the requirements, and the design is validated to ensure it is desirable to the market representative.

A guide to the concept development stage is provided in Table 5.1.

Purpose

The purpose of the concept development stage is to create a concept for the product and evolve it to contain enough information to create basic estimates for cost, size, weight, and feasibility. The concept that is evolved should be the one deemed most capable of satisfying the market requirements while simultaneously being capable of development within budget and according to the development schedule.

In addition, the concept development stage lays the foundation for the subsystem engineering stage by broadly defining subsystems and their interfaces. The

© Springer Nature Switzerland AG 2020
C. A. Mattson, C. D. Sorensen, *Product Development*,
https://doi.org/10.1007/978-3-030-14899-7_5

Table 5.1: Summary of the concept development stage.

Concept Development: Create a concept for the product and evolve it to have enough information to create basic estimates of cost, size, weight, and feasibility. Also include subsystem definitions, interface definitions, and target values for performance.

	Required information	Typical artifacts	Checking criteria	Approval criteria
Requirements	Subsystem requirements	Section A-D of the requirements matrix for each subsystem	See the Opportunity Development summary for the checking criteria for requirements matrices.	Complete and appropriate.
	Target values for the system and the subsystems.	Section E of the requirements matrix for the system and all subsystems.	Consistent with ideal values? Achievable with the selected product concept? Target values for all performance measures? Distinction between constraints, success measures, stretch goals? Trade-offs to less than ideal performance justified?	Consistent with expectations of market
Tests	Justification for concept selection, including model analysis and/or prototype test data demonstrating concept feasibility.	List, chart, or summary of considered concepts. Evaluation summary of considered concepts demonstrating feasibility of selected concept. Model and/or prototype test reports demonstrating the validity of the selected concept.	Are the test methods complete and correct? Is there enough detail for a third party to repeat the tests?	The desirability of the concept is demonstrated.
Models	Rough technical models of the product concept.	Low-fidelity fundamental models of the concept's operating principles. Statistical models describing the performance of experimental prototypes or related existing products.	Are the models reported in enough detail to allow a third party to use them?	Not directly approved; used with tests.
Prototypes	Simple prototypes of the product concept.	Rough prototypes (foamboard, paper, foam, clay, cardboard, plywood, etc.) used to show how the concept functions.	Are the prototypes appropriate for the intended use?	Not directly approved; used with tests.
Design	Geometric and other appropriate definition of the concept.	One or more of the following: Annotated hand sketches of concept. Overall system CAD model. Layout drawing. Skeleton drawing. Notes on sketches/drawings explaining concept. Block diagrams of electrical or fluid systems.	Is the design clearly communicated? Can a third party understand the intended design?	The design is sufficiently transferable to support cost, size, weight, and feasibility estimates.
	Decomposition of product concept into subsystems.	Tree or other relationship diagram that shows structure of decomposition. List of subsystems.	Are the subsystems and their relationships clearly shown?	The decomposition is appropriate for the selected concept.
	Subsystem interface definitions.	Interface matrix showing where subsystems interact. Product-focused requirement statements for each of the interfaces. Performance measures for each of the interfaces. Subteam responsibility assignments for each interface.	Is the interface matrix complete? Are the interface definitions complete? Are the interface definitions appropriate to achieve the desired performance?	The interfaces are fully defined and appropriate for the concept.
	Preliminary Bill of Materials.	Spreadsheet or database table that lists all known components, even if they have only a part name at this time.	Is the BOM complete at the level of known detail?	The bill of materials is appropriate.
Useful tools:		Bill of materials, bio-inspired design, brainstorming, competitive benchmarking, controlled convergence, decomposition, internet research, interviews, literature review, method 635, mind maps, prototyping, recombination table, requirements matrix, scoring matrix, screening matrix, sketching, theory of inventive problem solving (TRIZ).		
Common pitfalls:		Concept fixation; premature concept critique; reinventing the wheel; vague interfaces; decision delay.		

opportunity for each of the subsystems — meaning the requirements, performance measures, requirement-measure correlations, and ideal values — is also defined.

Finally, during this stage the trade-offs between the various market requirements are considered, and specific target values for performance measures are chosen.

Development Outcomes

Transferable artifacts containing the following information must be approved at the end of concept development.

Requirements

There are two primary elements in the requirements for this stage of development.

1. Subsystem requirements

Each subsystem has its requirements defined in the context of the entire system to ensure that the combination of subsystems will lead to desirable system performance. Like the system opportunity, the subsystem opportunity comprises requirements, performance measures, requirement-measure correlations, and ideal values. However, the opportunity is focused on the specific subsystem, rather than the system as a whole.

One important difference in the subsystem opportunity definition is the requirements. Rather than using market requirements, the subsystem opportunity definition uses selected system performance measures as the subsystem requirements, as shown in Figure 5.1. Subsystem performance measures are chosen to measure the achievement of the selected system performance measures. The relationships in the opportunity are relationships between the subsystem performance measures and the selected system performance measures.

2. Target values for the system and subsystems

Once an overall product concept is selected, the need for trade-offs between

the various performance measures will become more clear. It is rare that the concept will be able to achieve the ideal values for all of the performance measures. The team will often need to compromise in one area to achieve ideal values in other more important areas. Using rough technical models of the selected concept (often based on the performance of competing products), the team can choose target values for each of the performance measures. In all cases, the target values must be within the acceptable limits of the ideal values. In some cases, the ideal value will be chosen as the target. In other cases, due to trade-offs, the target value will be between the ideal and the acceptable limit.

For example, the human-powered water well drill team selected a concept similar to that seen in Figure 4.1; for this concept the deeper the borehole, the more substantial the structure needs to be so it can support the weight of a long drill string. Consequently, for this design concept, the deeper the borehole, the more expensive the structure to support it becomes. Therefore, when choosing a target value for borehole depth, the team may choose a depth that is different than the ideal value shown in Figure 4.2.

> *Can you think of a concept for which there is no tradeoff between maximum borehole depth and the cost to manufacture one drill?*

As the team chooses target values for performance measures, it can be helpful to recognize that peformance measures generally fall into four classes: basic requirements, constraints, key performance measures, and stretch goals. A clear understanding of the class of each performance measure can dramatically affect success in product development.

Target values for *basic requirements* are important only to the extent that they need to lie within the acceptable range. Basic requirements whose values differ from ideal have a minimal effect on the market's perception of the quality of the product. Therefore, it is generally not wise to use significant development effort to optimize these values.

Constraints can usually be recognized by yes/no performance measures. As long as the constraint is met there is not much concern about how well it is met. Therefore the target values for constraints are simply yes or no. For example, a car designed for the US must meet the Federal Motor Vehicle Highway Safety Standards in order to be a viable product.

Target values for *key performance measures* have a strong effect on the perceived quality of a product. For example, fuel economy is a key performance measure for an economy car. It represents one of the major criteria customers use to make purchase decisions. Hence, an economy car might have a target value for fuel economy of 45 miles per gallon.

Stretch goals are target values for key performance measures that would be amazing to reach, but are somewhat unlikely. In 2019, an electric car range of 400 miles between charges would be a stretch goal. It represents a target that has not yet been achieved, but if achieved it would dramatically increase the electric car market.

Development teams should choose the key performance measures that will make the biggest difference in market acceptance for the product. They should be sure they understand and meet the constraints. And they should have a limited number of stretch goals, in order to challenge the team but not distract from the key performance measures.

Tests

As part of the concept selection process, the development team must evaluate alternative concepts in terms of market requirements. Although these evaluation methods are generally high-level and somewhat qualitative, it is important that the critical methods be documented in a transferable way. Effective product

development requires more than just a guess as to the desirability of the concept.

The team must balance the need for transferability of the evaluation methods with the need to evolve the design quickly and efficiently. As a guideline, the evaluation methods should focus mostly on the selected concept, but provide sufficient rationale for its selection that someone outside the team would support the selection as appropriate.

During this stage, the concept should be tested using prediction or test methods to determine how well the concept is expected to meet market requirements.

Models

Rough technical models of the selected product concept are helpful in making trade-off decisions and choosing target values for performance. These models, which are often based on evaluation of competitive products in the marketplace, define expected relationships between the various performance measures for a given concept. For example, relationships exist between engine displacement, horsepower, and fuel economy for various classes of automobiles. These relationships can be used to help define the target values for a new automobile being designed.

Prototypes

Rough prototypes are often used in the concept development stage to communicate and explore how concept ideas function. Prototypes are very helpful to get the development team out of the virtual world and into the real world.

Teams should use prototypes, focusing on their use to evolve the design (definition of the product), rather than making the prototyping process or the prototypes themselves the purpose of the product development process.

Design

There are four required design outcomes for the concept development stage.

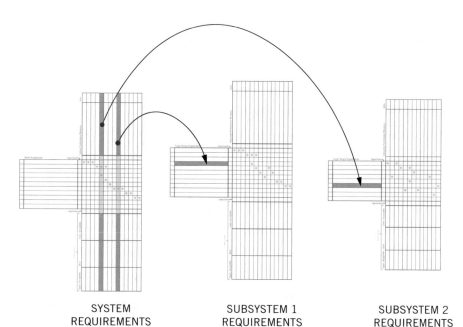

Figure 5.1: Subsystem requirements come from the target values of system performance measures. Not all system performance measures apply to every subsystem.

SYSTEM REQUIREMENTS   SUBSYSTEM 1 REQUIREMENTS   SUBSYSTEM 2 REQUIREMENTS

1. Definition of the concept

A product concept describes the means for achieving major market requirements, where enough detail is provided about the spatial and structural relationships of the principal components and subsystems that basic cost, size, weight, and feasibility estimates can be made[2]. The product concept must have a sufficient level of detail to be evaluated relative to the opportunity. The concept definition often takes the form of annotated sketches (manual or computer-generated) or a layout drawing. Along with the graphical representation of the concept, notes identify key aspects that are not readily understood from the drawing.

Where complex functionality is part of the selected concept, a block diagram is often included as part of the concept definition.

The concept definition must be clear and unambiguous enough to be understandable to a third party.

2. Decomposition of the concept

A formal definition of the subsystems included in the design permits independent work on each of the subsystems. This allows development subteams to work in parallel and accelerates product development. At this stage, the product definition includes a list of the subsystems; the subsystems will be developed fully during subsystem engineering.

It can also be helpful to show a tree or other representation of the logical relationships between the subsystems.

3. Subsystem interface definitions

The interfaces between subsystems are crucial constraints on the design for each subsystem. The definition of the interfaces includes an interface matrix that defines which interfaces exist, and an interface definition for each interface.

An interface definition consists of the following items (examples of each are given in Table 5.2):

---

[2]This definition is based on that provided by Dym and Little (2008).

- A description of the function that the interface must provide to each of the components or subsystems involved with the interface. This could be expressed in the form of product-focused requirement statements (see Product-Focused Requirement Statements (11.48) in the Development Reference) for the interface.
- Performance measures for the interface functions as available. For a complex interface, a full function-measure correlation matrix could be made, following the principles of a requirements matrix. For a simple interface, these performance measures can be correlated with a single interface function.
- Responsibility for each of the interface functions. In many cases, a subteam in charge of one interfacing subsystem will have primary responsibility for creating the interface definition, and the other subteam will simply use that definition in their design. In other cases, the interface definition will be created separately from any of the subteams developing the interfacing subsystems. Whichever way is chosen, the responsibility for defining the interface should be made clear.

The interface definition is often captured in an *Interface Control Drawing* (ICD), which is a formal specification of the interface. The subteam responsible for defining the interface is the owner of the interface control drawing. As with all engineering drawings, the ICD is placed under revision control. All subsystems using the interface refer to the ICD as part of the constraints on the subsystem design.

4. Bill of materials

At the end of concept development, major components and subassemblies are known. A bill of materials (see Bill of Materials (11.3) in the Development Reference) should be used to track all of the information about parts used in the design. In concept development, the bill of materials may only list the names of major parts. However, it forms a scaffold on which the final bill of materials can be built. It also helps plan the remainder of the development process.

Useful Tools

During the concept development stage one final system concept will be selected from a set of candidates. Therefore, the team will need to consider activities for both concept creation and concept selection. We recommend that the team consider the use of one or more of these activities for this stage of development. Most of these activities are described briefly in the Development Reference. Section 5.4 provides insight on how to develop a strong final system concept.

For generating concepts: Bio-Inspired Design (11.4), Brainstorming (11.5), Decomposition (11.14), Internet Research (11.33), Method 635 (11.35), Mind Maps (11.36), Recombination Table (11.53), Theory of Inventive Problem Solving (TRIZ) (11.66).

For selecting concepts: Controlled convergence (11.10), scoring matrix (11.59), screening matrix (11.60), multivoting (11.37).

For establishing target values: Benchmarking (11.2), interviews (11.34), literature review, internet research (11.33).

For general design work: Bill of materials (11.3), prototyping (11.50), requirements matrix (11.55), sketching (11.63), storyboards (11.64).

Common Pitfalls

- Concept fixation: Avoid fixation on one concept by thoughtfully exploring many concepts before converging. Identifying the best concept can only be done in the context of multiple concepts. Similarly, avoid fixation on one particular class of concepts. For example, avoid considering only human-powered water well drill

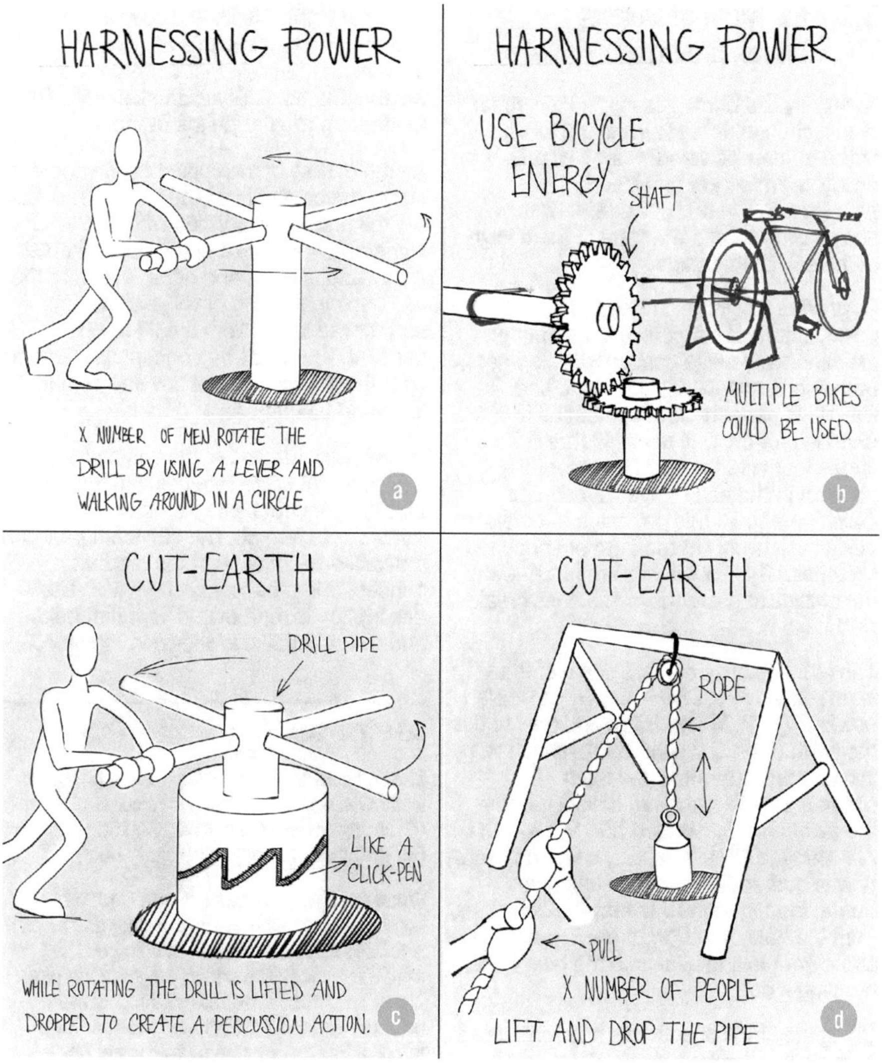

Figure 5.2: Early-stage concept sketches for human-powered water well drill.

concepts that remove earth using a spinning bit.

- Premature concept critique: Avoid critiquing ideas during concept generation activities. Infeasible ideas can lead to innovative feasible concepts.

- Reinventing the wheel: Avoid reinventing solutions that already exist; use existing, proven, technologies when they are available and suitable.

- Vague interfaces: Be sure to thoroughly define and control the interface definitions. Without well-defined interfaces, subsystem teams cannot work effectively in parallel.

- Decision delay: Set and stick to aggressive deadlines. Only consider allowing schedule slippage when specific deficiencies are described and a plan to overcome them is identified.

5.2   Example of Concept Development

To solidify the principles introduced so far in this chapter, let's return to the example of the human-powered water well drill, the details of which are provided in Appendix B. We first consider how the design evolved and the state of the design at the end of this stage.

To provide insight regarding the development of the concept, we briefly examine early-stage concept development, mid-stage concept development, and late-stage concept development as it unfolded for the drill team. Figure 5.2 shows four sketches that represent the concept in the early stages of concept development. At this point, the transferable design exists only in the form of annotated sketches. These sketches are just a few of the dozens and dozens of sketches created by the team.

The four sketches capture some of the team's ideas about how to harness human power (Figs. 5.2a and 5.2b) and how to cut the earth (Figs. 5.2c and 5.2d), which were two of many sub-areas the team considered. Although not described here, the team also considered how to allow the drill string to simultaneously spin and drop, how to add additional pipe segments during the dig, how to retract the drill string from the hole, and how to prevent the drill string from falling back into the hole while bringing it out.

Using the kinds of activities and tools listed in Table 5.1 the team evolved the early-stage concepts into those shown in Figure 5.3. We can see in Figs. 5.3a – 5.3c that the team converged on using a horizontal bicycle-driven shaft connected to the drill string by way of a bevel gearset. At this point in the development process, the team was concerned that the cost of the bevel gearbox may make this concept infeasible, so the team selected a second concept as a backup (Figure 5.3d).

The details in the renderings, though still somewhat abstract, are about the main spatial and structural relationships of the

principal components and subsystems. To that end, we can see that the design is evolving to the point that basic cost, size, weight, and feasibility estimates could be made from the design information.

By the middle of the concept development stage, the concept has indeed evolved, but the team was not satisfied with the vulnerability and cost of the bevel gearset. After various tests, and other research, the team eliminated the bevel gear concept, and moved ahead with the back-up concept. The back-up concept as it existed near the end of concept development is shown in Figure 5.4a and 5.4b.

Importantly, the end of the concept development stage requires that the product concept be captured in a transferable format. The drill team provided a detailed description of the product concept and a concept schematic. The description is reproduced verbatim below, and the schematic is shown in Figure 5.5.

SELECTED CONCEPT

The selected concept for the human-powered water well drill consists of five major subsystems: (i) turning wheel, (ii) kelly bar, (iii) support tables, (iv) structural frame, and (v) slurry pump.

When operated, the drill cuts the earth with a commercial six-inch tri-cone or drag drill bit as it is rotated up to 30 RPM, depending on the required torque. The added weight of the pipe (up to 1,800 lb)[3] provides natural downward pressure to grind through hard rock. A pump forces a thick slurry (made from water and a sealant called bentonite) down the pipe and back up around it, both carrying out the cuttings and sealing the borehole walls with the added bentonite. To add and remove additional pipe, an eight-foot structure is set into place that facilitates the use of pulleys and clamps. The selected concept is an integration of professional and homemade systems, as detailed below.

The wheel consists of eight spokes that come out of a steel plate in the middle, as shown in Figure 5.5. The spokes are one-inch square

[3]As shown in an appendix provided by the team, but not reproduced in this book.

Figure 5.3: Mid-stage concept renderings for human-powered water well drill.

steel tubing providing strength as well as ease in welding them to the plate. The spokes are connected at the outer ends by more steel tubing, creating a circular wheel six-feet in diameter with twelve evenly spaced pegs around the outside. This diameter and number of pegs allows for several people to operate, or turn, the drill comfortably at a sufficient RPM without having to walk around in circles, based on preliminary testing.

The kelly bar is a five-foot long section of four-inch square tubing that has a section of pipe welded inside it. This makes it possible to screw into the other sections of pipe. The kelly bar fits through a square hole in the middle of the steel plate of the wheel, allowing it to turn as the wheel is turned, thus turning the pipe as well. While the wheel rests on a simple thrust bearing on top of the table to minimize friction, the kelly bar and pipe move downward through the square hole as the drill cuts the earth. This allows the drill to be turned and dropped simultaneously in a very simple manner without gears or complex components.

Figure 5.4: Late-stage concept renderings for human-powered water well drill.

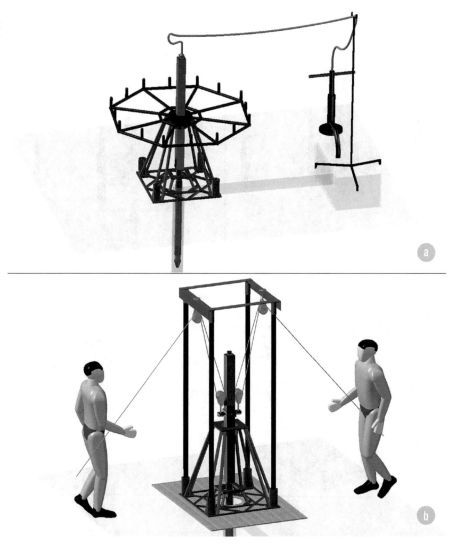

The abovementioned table is at a height of three feet, supported by eight braces which are connected to a three-foot square base plate on the ground. Based on a standard human height and comfortable range of motion, this height is ergonomically optimal for the operators, and also allows the structure to be short enough for mobility, structural integrity, and the desired low cost. Another, smaller, table is connected to the base plate at a height of eight inches to support the clamps that hold the weight of the pipe during its removal. To reduce cost and weight while maintaining necessary strength, the base plate is made up of a web of steel strips instead of a solid plate, as seen in Figure 5.5.

When the pipe needs to be lifted, the wheel is removed, and the structural frame is attached to the base plate. The cylindrical supports of the frame fit into hollow, cylindrical sections sticking out of the corners of the base plate. The top of the structure is held together by a single steel piece with hollow cylindrical sections that slip over the top of the supports. This design makes the structure extremely simple to assemble and disassemble while still being sturdy. The

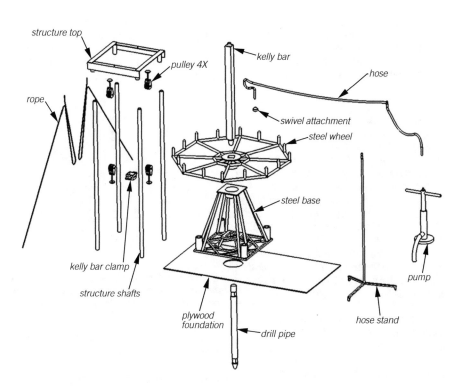

structure incorporates a pulley system in which two or more people pull in opposite directions in order to lift out the thousands of pounds of pipe with relative ease. The combined impact of the pulleys will reduce the required lifting force to 1/16 of the actual weight of the pipe. Therefore, the lifting burden for each individual is at maximum 115 pounds (based on 2 operators).

Once a five-foot length of pipe has been drilled into the ground, the entire drill pipe assembly is lifted up and clamped in place. The kelly bar is removed from the top pipe, a new pipe is added on top, and the kelly bar is attached to the new pipe. When the pipe is again lowered into the borehole, the kelly bar will have its full length to travel down through the square hole as it rotates. The process to remove the pipe is simply the reverse, and the pipe is continuously pulled up, rather than up and down. It is estimated that if the device drills eight feet each day (as suggested by the client), this process of adding a new pipe will only occur at most twice a day.

The final component of the selected concept is a slurry pump that is used to pump the re-circulated water-bentonite slurry from a

nearby trough back down the center of the pipe to lift out the cuttings as the drill digs through dirt and rock. A treadle pump is selected because it utilizes the operator's body weight to create the necessary pressure and volumetric flow rate down the pipe, and is based on common treadle pump designs that are currently made and used in developing countries.

The system described above is a combination of the professional drilling technology with the manual and inexpensive nature of homemade drilling systems. Designed for developing countries, it utilizes a nearly tool-less assembly and relies heavily on welding which is virtually available all over the world. Due to the trade-off between torque and angular velocity and the average capacity of human power output, the drill is capable of supplying up to 30 RPM at 150 ft-lb. or up to 1,500 ft-lb. at 2.5 RPM. Depending on the attributes of the current soil or rock formation, the operators can easily adapt the input to continue the drilling process[4]. The

_____
[4]The team also provided more details and graphs on horsepower requirements in an appendix.

selected concept is a product of months of iterative concept ideation, modeling, analyzing, and selection, all centered on the market requirements and performance measures. The selected concept is a powerful solution to the client's needs.

> *Check the design. Does the above concept description together with the concept schematic (Figure 5.5) and renderings (Figure 5.4) unambiguously define the spatial and structural relationships of the principal components and subsystems? Has a justification been provided that describes why the selected concept has strong potential to meet the opportunity? Is the concept transferable; could it be transferred without confusion to another team for further development? How would you improve the desirability or transferability of the design?*

In addition to the product concept, a decomposition is required as part of the system architecture. The team chose to structurally decompose the drill into the following major subsystems, as shown in Figure 5.7:

- Drill string — includes kelly bar, drill pipe, drill bit

- Drill string support and lift structure (hoist) — includes the structure that can support the weight of the pipe, and the mechanism for providing humans with mechanical advantage to lift and lower the pipe with relative ease

- Wheel — includes just the wheel

- Wheel support — includes whatever keeps the wheel at an optimal height

- Pump system — includes the pump, hoses, filters, pumping fluid, and chip separation scheme

To facilitate the independent development of each subsystem, the interface requirements must also be established and adhered to by the developers of each subsystem. To do this, the team created an interface matrix, shown in Figure 5.6. The center portion of the matrix shows which subsystems have defined interfaces with other subsystems. These are represented by the dot. For each dot in the matrix, an interface definition is created.

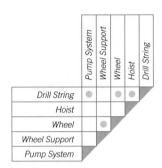

Figure 5.6: Interface matrix for drill. This matrix shows which subsystems have interface requirements with other subsystems.

Table 5.2 shows a sampling of the interface definitions for the drill.

With the decomposition in hand, the team defined the subsystem opportunities in a set of subsystem requirements matrices. To illustrate how this was done, we've provided three subsystem requirements matrices for the drill. They are shown in Figs. 5.8–5.10. To create these matrices, the team started by examining the list of target performance measures from the system requirements matrix and determined which ones applied to each subsystem. They then created a requirements matrix for each subsystem, where the applicable system target performance[5] was listed as a requirement of the subsystem matrix[6] as shown in Figure 5.8. The same procedure used to determine performance measures for the system was used to develop subsystem performance measures for each system target. Units, ideal values, and target values are chosen for each subsystem performance measure.

---

[5]Sections B and D of system-level requirements matrix.

[6]Section A of the subsystem level requirements matrix.

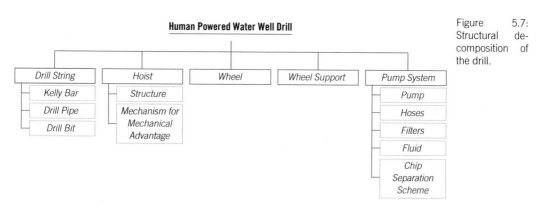

Table 5.2: A sample of the interface definition for the drill. Although the team did not create them, it would have been helpful to have an interface control drawing for each of the interfaces defined in this table.

Interface Between	Interface Functions	Performance Measures	Responsibility
Drill string and pump system	Maintain pressure	Pressure drop < 1 psi	Pump system team
	Accommodate flow	Flow rate 30 – 50 gpm	Pump system team
	Attach to drill string	2 inch female NPT	Drill string team
	Allow relative rotation between string and pump	Up to 30 RPM, < 2 lbf-ft of torque	Pump system team
	Attach to pump system	1.5 inch female NH or NST	Pump system team
Drill string and wheel	Transmit torque	Up to 1,500 lbf-ft	Drill string team
	Allow axial motion	Maximum of 10 lbf of force	Drill string team
Drill string and hoist	Support drill string weight	Up to 1800 lbf	Drill string team
	Connect to hoist, fit with hoist hook	Opening diameter ≥ 1.5 inches	Hoist team
Wheel and wheel support	Minimize torque loss	< 2 lb-ft at 30 RPM	Wheel support team
	Maintain drill string location	Center moves < 0.5 inches during use	Wheel support team

The set of requirements matrices (including any updated/refined requirements matrices) represent the evolved state of the requirements. During the concept development stage the system requirements matrix also evolves. New market requirements may be identified, ideal values may be refined, and more. At a minimum, target values are added for each performance measure after the final concept is selected. The target values for the drill system are shown in Table 7.2.

Because of space limitations, they are not shown in a requirements matrix, but they were added to the requirements matrix after the final concept was selected during the concept development stage.

Both the design and the requirements were checked and tested by the team, and approved by the client. The transferable nature of the concept description, the concept schematic, and the requirements matrix facilitated the approval process.

Product: DRILL
Subsystem: STRUCTURE

Subsystem Performance Measures	#	Units
Maximum stress in structure at twice the borehole depth	1	psi
Rating of mechanism used to lift twice the full pipe weight	2	lbs
Maximum stress in structure during drill pipe change over	3	psi
Downward drilling force	4	lbs
Rate of drill string decent	5	ft/min
Weight of heaviest structure welded sub-assembly	6	lbs
Length of longerst structure after development	7	in
Percentage of structure manufacturable in Tanzania	8	%
Percentage of mechanical advantage mechanism manufacturable in Tanzania	9	%
Cost to produce structure after development	10	USD
Number of parts in structure subsystem	11	
Time required to learn to set up the structure	12	min
Time required to learn to attach an auxilliary subsystem	13	min
Structure is attractive	14	n/a
Structure interest investors	15	n/a

#	Target Design Requirements	Importance	1	2	3	4	5	6	7	8	9	10	11	12	13	14	15
1	Borehole depth of 220 feet	9	●	●	●												
2	Time 60 minutes to drill though 6 inches of rock	10				●	●	●									
3	Downward drilling force of 500 lbs	3				●											
4	36 feet depth cut per 8 hours drilling	9					●										
5	Weight of heaviest assembly less than 200 lbs	9						●									
6	Longest dimension less than 96 inches	9							●								
7	90% of drill manufacturable in Tanzania	9								●	●						
8	$1500 USD produce 1 drill after development	9										●	●				
9	Less than 8 hours to learn how to use the drill	9												●	●		
10	Drill is attractive	4														●	
11	Drill interests investors	12															●
	Importance		9	9	9	10	19	19	9	9	9	9	9	9	9	4	12

Ideal Values	1	2	3	4	5	6	7	8	9	10	11	12	13	14	15
Upper Acceptable	50,000	–	50,000	–	–	200	96	–	–	1,000	100	240	240		–
Ideal	25,000	4,500	25,000	3,000	4	50	84	100	100	750	30	60	60	Drill has a professional look. People are interested in looking at it.	Drill captures media attention. Investors are proactive in contributing. Drill has iconic look.
Lower Acceptable	–	2,250	–	500	1	–	–	90	0	–	–	–	–	Drill looks like a piece of machinery for drilling holes.	The drill, when explained to investors, is something they want to invest in.

Figure 5.8: Subsystem requirements matrix for structure subsystem.

Product: DRILL
Subsystem: WHEEL

Subsystem Performance Measures (with Units):

1. Spinning speed — RPM
2. Wheel diameter — ft
3. Maximum stress in wheel with 600 ft-lbs torque applied — psi
4. Maximum stress in handle with 600 ft-lbs torque applied — psi
5. Distance between handles — deg
6. Weight of heaviest welded wheel assembly or component — lbs
7. Longest dimension of welded wheel assembly — ft
8. Percentage of wheel manufacturable in Tanzania — %
9. Cost to produce one wheel after development — USD
10. Time required to learn to put wheel tolerance — min
11. Time required to learn to add the wheel to the support — min
12. Time required to learn how to spin the wheel efficiently — min
13. Height of lowest point to grab handle (center of palm) — in
14. Height of highest point to grab handle (center of palm) — in
15. Feels comfortable — n/a
16. Wheel is attractive — n/a
17. Wheel interests investors — n/a

#	Target Design Requirements	Importance	1	2	3	4	5	6	7	8	9	10	11	12	13	14	15	16	17
1	Time 60 minutes to drill through 6 inches of rock	10	●																
2	300 ft-lbs applied torque to the drill string	10		●	●	●													
3	36 feel depth cut per 8 hours drilling	3	●																
4	4 people maximum required to use the drill	9		●			●												
5	Weight of heaviest assembly less than 200 lbs	9						●											
6	Longest dimension less than 96 inches	9							●										
7	90% of drill manufacturable in Tanzania	9								●									
8	$1500 USD produce 1 drill after development	9									●								
9	Less than 8 hours to learn how to use the drill	9										●	●	●					
10	Height of hand operated parts to be between 2 and 5 feet	1													●	●			
11	Feels comfortable	1													●	●	●		
12	Drill is attractive	4																●	
13	Drill interests investors	12																	●
	Importance		13	19	10	10	9	9	9	9	9	9	9	9	2	2	1	4	12

Ideal Values:

	1	2	3	4	5	6	7	8	9	10	11	12	13	14	15	16	17
Upper Acceptable	–	–	50,000	50,000	200	96	–	300	240	240	240	–	–	48	–	–	–
Ideal	30	8	25,000	25,000	45	50	48	100	100	60	60	60	41	44	Drill can be operated continuously without the need to rest. Does not require awkward movements.	Drill has a professional look. People are interested in looking at it.	Drill captures media attention. Investors are proactive in contributing. Drill has iconic look.
Lower Acceptable	3	6	–	–	30	–	–	90	–	–	–	–	39	–	Drill can be operated with occasional rest, and requires awkward movements that leave the user sore.	Drill looks like a piece of machinery for drilling holes.	The drill, when explained to investors, is something they want to invest in.

Figure 5.9: Subsystem requirements matrix for wheel subsystem.

Product: DRILL
Subsystem: WHEEL SUPPORT

	Target Design Requirements	Importance	1 Size of access area to borehole (ft^2)	2 Weight of wheel support (lbs)	3 Longest dimension of wheel support (in)	4 Percentage of wheel support manufacturable in Tanzania (%)	5 Cost to produce the wheel support and connections (USD)	6 Time required to learn to put wheel support in place (min)	7 Time required to llearn to remove the wheel from the support (min)	8 Time required to learn to add the wheel support to the structure (min)	9 Height of wheel support, where wheel is supported (ft)	10 Wheel support is attractive (n/a)	11 Wheel support style matches the rest of the drill (n/a)	12 Wheel support makes the drill interesting to investors (n/a)
1	36 feet depth cut per 8 hours drilling	3	●											
2	Weight of heaviest assembly less than 200 lbs	9		●										
3	Longest dimension less than 96 inches	9			●									
4	90% of drill manufacturable in Tanzania	9				●								
5	$1500 USD produce 1 drill after development	9					●							
6	Less than 8 hours to learn how to use the drill	9	●					●	●	●				
7	Height of hand operated parts to be between 2 and 5 feet	1									●			
8	Drill is attractive	4										●	●	
9	Drill interests investors	12												●
	Importance		12	9	9	9	9	9	9	9	1	4	4	12
	Lower Acceptable		1.50	–	–	80	–	–	–	–	1.5	Drill can be operated with occasional rest, and requires awkward movements that leave the user sore.	Drill looks like a piece of machinery for drilling holes.	The drill, when explained to investors, is something they want to invest in.
	Ideal		16	25	36	100	100	30	30	30	3.5	Drill can be operated continuously without the need to rest. Does not require awkward movements.	Drill has a professional look. People are interested in looking at it.	Drill captures media attention. Investors are proactive in contributing. Drill has iconic look.
	Upper Acceptable		–	100	96	–	200	240	240	240	-1.5	–	–	–

Figure 5.10: Subsystem requirements matrix for wheel support subsystem.

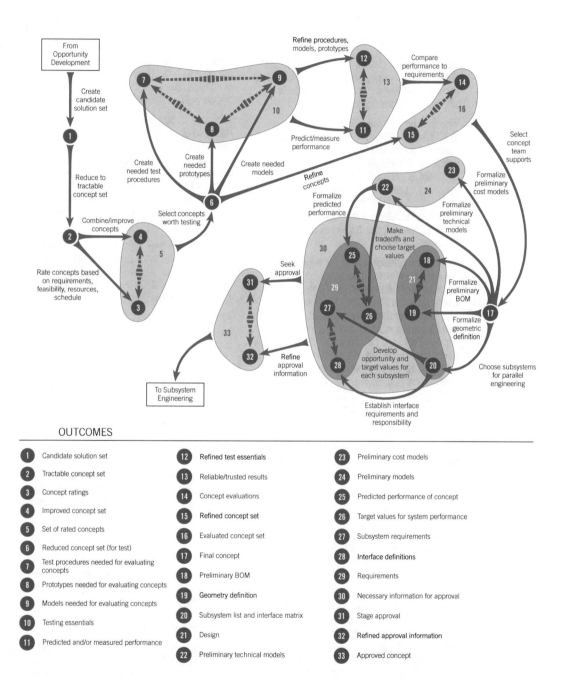

Figure 5.11: Top-level activity map for the concept development stage.

5.3   Top-Level Activity Map for Concept Development

Starting with the approved opportunity from the previous stage, a top-level activity map for the concept development stage is shown in Figure 5.11.

The first half of the diagram (Outcomes 1-17) focuses on developing a strong concept (Outcome 17) for the overall product. Converging on a strong product concept is perhaps one of the most important parts of the product development process as it determines what will be involved in finishing the design. Because it is so important, the top-level activity map is relatively detailed.

Concept development begins with generating a candidate solution set (Outcome 1) and refining through evaluating and reducing the set to a tractable number (Outcome 2). Concepts are rated, combined, and improved to arrive at a rated concept set (Outcome 5). From this rated set, a smaller set is selected for testing with models and prototypes (Outcome 10). The models and prototypes are used to predict and measure performance of the concept, which is compared with the requirements, and used to further refine the concepts (Outcome 16). Finally, the team selects a concept to carry forward to further development (Outcome 17).

Note that a team will develop strong concepts many times during product development, including concepts for subsystems. A more generic process, along with some key methods, for developing strong concepts, is discussed in Section 5.4.

Once a concept has been selected, it is time to formalize the required information for approval. This includes preliminary technical and cost models (Outcomes 22 and 23) that can be used to establish trade-offs for selecting target values (Outcome 26) and provide preliminary predictions for key performance measures (Outcome 25). We also formalize a preliminary Bill of Materials (Outcome 18), the geometric and/or logical design of the system (Outcome 19), and a selection of subsystems to be engineered in parallel (Outcome 20). These three outcomes constitute the design for concept development (Outcome 21).

With the subsystems identified, we are ready to complete the requirements by establishing the interface definitions (Outcome 28) and the subsystem target values (Outcome 27). Together with system target values and predictions of performance, these constitute the requirement information for the concept development stage.

With the approval information complete, stage approval can be sought. Note that stage approval may be an iterative process, requiring updates to any or all of the approval information.

Although this activity map has many details, we can recognize that it is high-level because most of the activities in the map are general activities. In order to use this map on a project, the general activities must be made more detailed and specific to a unique project by at least one level.

5.4   How to Develop a Strong Concept

At virtually any stage of the product development process, when we need an optimal way to meet a critical design need, we are likely to use the pattern of developing a strong concept. The purpose of this section is to show how to develop a strong concept to meet a design need.

A *concept* is an idea for solving a design need that includes a description of the main operating principles, major components, and structural, logical, and operating interfaces between the components.

A *strong* concept is a concept that has been rated as superior to any other known concept, with specific advantages and few or no disadvantages. It has been selected as a concept that is more likely than others to lead to desirable performance.

There are five fundamental steps in developing a strong concept. These steps are shown in Figure 5.12. The first is to create a quality set of potential concepts (using creating activities). The second is to reduce the relatively large set of concepts to a smaller set that can be analyzed in more detail (using evaluating and deciding activities). The third and fourth steps result in interdependent outcomes. The third step is to rate each remaining concept according to how well it meets the design needs (using evaluating activities), and the fourth step is to combine highly rated concepts or parts of concepts with others to create improved concepts (using creating activities). The interdependence of the outcomes means that we will ideally continue to rate and improve the set of candidates until no better concepts are found. The final step is to select the best concept (using deciding activities).

A. Create Candidate Solution Set

One of the challenges of product development is determining what the best possible design is. It's important to recognize that the *best* design can only be identified in the context of *multiple* designs. For this reason it's essential that we create a set of potential solutions, rather than just a single solution.

Concept set quality: Given that we can only do comparative rating of concepts, it is important to have a good set of concepts with which to make comparisons. It's not saying much to affirm that a concept is the best of two alternatives. It's somewhat better to say that it's the best of ten alternatives. It's better still to say that the concept is the best of fifty alternatives that have spanned the design space, including some that have pushed the boundaries of current knowledge.

If the rating of the concepts can be no better than the quality of the concepts against which they are rated, it is important to evaluate the appropriateness of a

concept set. Shah et al. (2003)[7] describe four attributes of a concept set that can be evaluated to assess its appropriateness: concept quantity, concept variety, concept novelty, and concept quality. Three of these attributes are illustrated in Figure 5.13.

Ultimately we want to know that we have considered a sufficient number of candidates of sufficient quality, variety, and novelty. Critical parts of the design should have large number of candidates with high quality, variety, and novelty in the concept set. Mundane parts, however, can have lower quantity, less variety, less novelty, and acceptable quality in the concept set.

In order to evaluate variety and novelty, it is useful to classify concepts along one or more design axes. These axes represent different dimensions that could be chosen to capture general characteristics that differ between the concepts. There is no "right" set of axes to use for a given design. It is sufficient to note that as a team thinks carefully about their design, it is likely that axes will become apparent. Shah et al. (2003) propose that often there will be axes that include physical principles (weight, length) and working principles (combustion engine, electric motor).

One of the tools used to classify concepts is a concept classification tree[8]. A concept classification tree graphically presents the different branches explored when generating alternative solutions or concepts. In addition to providing a measure of variety, the concept classification tree may provide additional ideas by highlighting branches that are not sufficiently explored.

A concept classification tree used by the human-powered drill team is shown in Figure 5.14. Note that although the team had as their primary focus the creation of a human-powered drill, during concept

---

[7]This paper also includes a quantitative method for evaluating novelty, variety, and quality. As of July 2013, the paper is available on-line at http://w.ecologylab.net/research/publications/ShahVargas-HernandezSmith2002.pdf

[8]See Concept Classification Tree in the Development Reference.

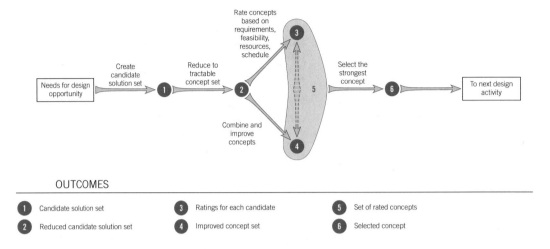

OUTCOMES

1 Candidate solution set   3 Ratings for each candidate   5 Set of rated concepts

2 Reduced candidate solution set 4 Improved concept set   6 Selected concept

Figure 5.12: Typical activity map for developing a strong concept.

Figure 5.13: Factors determining solution set adequacy. Black lines represent the borders of the known design space and dots represent candidate solutions. Candidate set quantity is the number of distinct candidate solutions. Variety is an indication of how broadly the design space has been explored. Novelty is a measure of how the design space has been expanded. Although not illustrated here, quality is an evaluation of how well the candidate solutions meet the needs of the design.

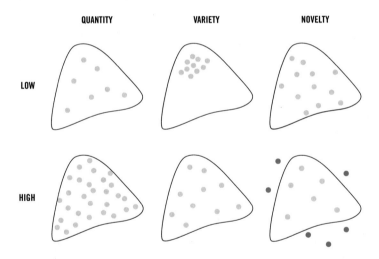

creation they expanded their thinking to consider solar power as a potential energy source. Perhaps their solution set would have been even better had they considered wind and rain as potential energy sources as well. By placing the concepts in a tree that shows logical relationships, we can see areas that could profitably be explored more fully.

Creating the solution set   A solution set should be developed using the methods similar to those described in Section 3.3. The solution set should have appropriate quality, quantity, variety, and novelty, as described above.

Deciding when the solution set is sufficient requires the use of judgment. Certainly, the more critical the design problem is to the

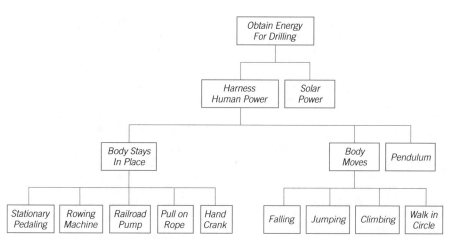

Figure 5.14: A concept classification tree for the harness energy sub-function of the human-powered drill.

overall product, the more challenging it is to converge on a good solution, as higher levels of quantity, quality, variety, and novelty in the solution set will likely be needed. Short-circuiting the solution set creation process may lead to missing an ideal solution. Prolonging the solution set creation process may put the overall development project behind schedule.

Once a solution set is deemed appropriate, reduction of the solution set is undertaken.

B. Reduce the Candidate Solution Set

If an excellent job of solution creation has been done, there will likely be too many concepts to thoroughly evaluate all of them. Therefore, some form of intuition-based evaluation is generally done to select approximately 5-15 concepts from the larger solution set. One common method of eliminating less-promising candidates is multivoting (Multivoting (11.37) in the Development Reference).

Once the solution set has been pared to a reasonable size, a more formal and thorough evaluation can be undertaken.

C. Rate the Concepts

When the potential solution set has been reduced to a tractable number of concepts, the concepts are placed in a screening

matrix[9] or a Pugh concept selection matrix[10]. The market requirements[11] are listed as labels for the matrix rows. The concepts under consideration are listed as labels for the matrix column, with a reference or benchmark concept listed in the first column. The choice of the reference concept is up to the team. It is often the lowest-risk or least novel concept in the set.

With the market requirements and the concepts in place, ratings are then given to each of the concepts for each of the market requirements. First, the column of the reference concept is filled with "=," meaning "the same as the reference." Then, working row by row, each concept is compared with the reference concept to see how well it will meet the market requirement. The concept will be given a "+," "=," or "-" if it is better than the reference, the same as the reference, or worse than the reference at meeting the market requirement.

Although it may seem like a small detail, it is important to move row by row, rather than column by column, through this evaluation process. Two important

---

[9]See Screening Matrix (11.60) in the Development Reference for more information

[10]See Controlled Convergence (11.10) in the Development Reference

[11]Section A of the requirements matrix.

objectives are met by moving row by row. First, it is much easier to be consistent in the evaluation of a particular market requirement if all the concepts are evaluated at the same time. Second, moving requirement by requirement helps prevent gamesmanship in artificially raising or lowering scores for specific concepts.

After all of the market requirements have been considered, the ratings of each concept are summarized at the bottom of the matrix. The number of pluses, minuses, and equals in each column is counted, and a net score is calculated as the difference between the number of pluses and the number of minuses.

D. Combine and Improve Candidates

If the ratings indicate that one concept is clearly superior, it may be selected. In most cases, however, there will be no single concept that is clearly superior. Each concept will have its strengths and weaknesses. The next step is to try to combine concepts with complementary strengths to see if a combined concept can be created that is better than either of the original concepts. The lowest-rated concepts are then removed from the matrix, the new concepts added, and the rating process repeated.

As the number of concepts is reduced through eliminating weaker concepts, the level of detail for both the concepts and the evaluation is increased. Some practitioners recommend that for final concept selection a concept scoring matrix is used. This chapter does not describe the use of a concept scoring matrix, but the Development Reference does.

Figure 5.15 shows an intermediate concept screening matrix for the harness energy subconcepts for the human-powered well team. Note that some concepts are rated poorly and scheduled for elimination, while others are scheduled for combination or further research and evaluation.

E. Choose the Best Concept

Having evaluated all the concepts relative to the market requirements, the team must

now choose the best concept. The best concept is not necessarily the one that has the highest score from the screening matrix. In fact, it is possible that several concepts will have equal scores. In addition to the market requirements, there are likely to be other criteria that may be important to the team. These other subjective requirements should also be considered.

When the best concept is selected, the team should be able to identify why the concept is superior without referencing the score on the screening matrix. The strengths of the concept should be clear, and should be easily listable by the team.

Having selected a best concept, the concept should be evaluated to see if the team really believes it is strong enough to proceed to further development. If so, the concept creation activity is complete. If not, the concept creation activity should be iterated in order to develop a stronger concept.

5.5   How to Generate Concepts

Generating concepts for products can be one of the most rewarding parts of product development. There are two general methods of generating concepts: exploring existing solutions and creating new ideas.

Existing Solutions

The world is exploding with products to meet the needs of billions of people. Ignoring the existence of these solutions when trying to solve design challenges almost certainly guarantees the development of products with one or more substantial flaws. At the least, it guarantees that development will take unnecessary time and resources.

Existing solutions are helpful to a development team in at least two ways. First, an existing solution may be able to be used as-is or with minimal modification as a component for a newly developed product. Second, an existing solution may be used as the basis for a new design.

In order to use existing solutions, relevant solutions must be found. We present three

Market Requirements	Rota-Sludge (Benchmark in Tanzania)	People Walking in Circle	Tugging with Rope	Stationay Pedaling	Rowing Machine	Railroad Pump	Climbing or Falling	Pendulum	Hand Crank	Solar Power
Torque	=	+	+	+	+	+	+	+	+	=
Speed	=	+	+	+	+	+	+	=	+	+
No. People	=	−	−	−	−	+	−	+	+	+
Simplicity - Device	=	+	+	−	−	−	−	−	−	−
Simplicity - Use	=	+	+	+	+	+	+	+	+	+
Manufacturability	=	+	+	−	−	−	−	−	−	−
Stationary	=	−	−	+	+	=	−	=	=	+
Body Stress	=	−	−	−	−	−	−	+	−	+
Size	=	+	+	=	=	−	−	−	=	=
Number of +'s	0	6	6	4	4	4	3	4	4	5
Number of ='s	9	0	0	1	1	1	0	2	2	2
Number of −'s	0	3	3.	4	4	4	6	3	3	2
Net Score	0	3	3	0	0	0	−3	1	1	3
Decision	—	Keep	Combine	Keep	Keep Research	Combine	Discard	Discard	Discard	Discard

Figure 5.15: Concept screening matrix for harness energy subconcept for human-powered drill. This matrix was created in the middle of the concept development stage.

commonly used methods for finding existing solutions, each of which is explained more fully in the Development Reference.

Internet Search

Perhaps the most commonly used method of finding existing solutions is a search of the internet. Internet search engines can return millions of hits for a given topic. One of the greatest challenges can be separating the desired information from the noise.

As an example of the value of an internet search, a project was recently proposed to develop an improved portable dental chair by modifying an existing chair. A Google search of "portable dental chair" returned 806,000 results. On the first page of the results there were images of at least seven different commercially available chairs. There were three different sellers of multiple chairs. There was an article written by a non-profit world dental organization that described the strengths and weaknesses of various commercial chairs. In less than five minutes, the searcher obtained a good understanding of the existing solutions.

One potential problem with internet searching is the assumption that all the good information is on the first few pages returned. In some cases, later pages have the best solutions. Deciding when it's time to end a search requires good judgment.

Catalog Search

Catalogs are another source of existing solution ideas. Catalogs are available both online and in paper copies. Paper catalogs are often easier to browse through than electronic catalogs, while electronic catalogs are often easier to search for specific items.

Catalogs from specific manufacturers, such as Timken for bearings or Suspa for gas springs, generally contain the most specific information and are very helpful for details when a concept has been chosen.

Catalogs from distributors, such as MSC or McMaster Carr, contain a wide range of products and may be useful for browsing just to see what is available.

Catalogs for consumer, rather than industrial, products are often very helpful in finding ideas that can be used as the basis for a design solution, even though the product itself is unlikely to be used.

Delphi Method

The Delphi method is based on interpersonal networking. The person looking for an existing solution finds a few people who might have ideas about such a solution. Each of these people is contacted, and asked about solution ideas. At the end of the interview, the person is also asked for the contact information of other people who might have solution ideas.

The Delphi method is helpful because you are drawing on the expertise of others without making large demands on their time. Using only a few minutes from any one person, the team is able to get access to a broad collection of expertise. The results from the Delphi search are often used as a starting point for internet or catalog searches.

Design Benchmarking

An effective method of evaluating competitive products is design benchmarking. In design benchmarking, the team discovers how competitive products have been designed to meet the market requirements. This not only provides the development team with great ideas, it also sets a target for the team to beat.

To perform design benchmarking, the development team purchases competing products and disassembles them. Products are analyzed to determine how each of the functions are delivered. Estimates of the manufacturing cost and reliability are made for each of the parts and assemblies. The disassembled products are often attached to visible display boards.

The designs judged best in terms of quality, manufacturability, and reliability become the standards against which all of the team's design ideas will be judged. This helps prevent the creation of inferior designs.

One world-class development organization required development teams to perform competitive benchmarking for every project, with a strict injunction that if the team could not develop a superior design,

it would have to use the competitor's design. This provided a great source of motivation, inspiration, and creativity.

Creating New Ideas

There are many activities for creating candidate solutions to design problems. Brainstorming is undoubtedly the most often referred to creation activity[12]. While brainstorming is generally useful throughout the product development process, other methods can also be useful. A list of typical creation activities for each of the stages of development is shown in Table 5.3. More information about most of these activities can be found in the Development Reference.

The Delphi method was described previously as a method for discovering existing solutions. If the persons being interviewed are asked about ideas they might have for *new* solutions, rather than existing solutions, the Delphi method becomes a method for creating candidate solutions.

Decomposition and Recombination

Often a design problem is sufficiently complicated to require concepts in multiple areas in order to define an overall solution. In this case, an effective tool for helping with concept creation is decomposition and recombination.

When using this tool, the overall product is broken down into manageable pieces. Ulrich and Eppinger (2012, pp. 120–124) recommend functional decomposition, although they also describe decomposition by user actions and by market requirements.

Once the product is decomposed, idea generation occurs for each of the pieces. We term a concept for solving only one piece of the product a subconcept. Each subconcept set can be evaluated for

---

[12]Kelley and Littman (2001) provide seven rules for effective brainstorming. Also see Brainstorming (11.5) in the Development Reference for more information

Table 5.3: Typical creation activities used in various development stages.

Opportunity Development	Concept Development	Subsystem Engineering	System Refinement	Producibility Refinement	Post-release Refinement
Brainstorming Delphi Method	Brainstorming Bio-Inspired Designs Method 635 Gallery Method Synectics Recombination Tables Function Structures	Brainstorming Bio-Inspired Designs Synectics CAD QFD	Brainstorming Recombination Tables CAD	Brainstorming TRIZ	Brainstorming TRIZ

novelty, variety, and quantity[13]. It may be difficult to evaluate the quality of an individual subconcept since product performance depends on the integration of multiple subconcepts.

When subconcept sets have been developed for each of the pieces, complete concepts are developed by combining one of the subconcepts from each piece. It may be possible to get multiple concepts from one combination.

The total number of possible combinations is the product of the number of ideas in each of the sub-areas. This is not the same as the number of concepts developed, for two reasons. First, not all of the combinations are guaranteed to be feasible. Second, as mentioned earlier, there may be more than one concept for a given combination.

The human-powered drill team decomposed their problem into a number of areas. Two of the most important were harnessing energy and cutting earth. For each of these areas, the team developed ten ideas to evaluate thoroughly. The ten ideas in each of two areas provide 100 different combinations. Not all of the combinations were turned into concepts by the development team.

The use of decomposition and recombination in concept creation can greatly improve the candidate solution set.

5.6   Summary

During the concept development stage, the team formalizes the architecture of the product. This includes not only generating and selecting a concept, but also decomposing that concept into major components or subsystems with defined interfaces and performance expectations (as laid out in the opportunity definition). To do this, the team will need to both generate and evaluate candidate concepts. Concept generation and evaluation tools can help improve the quality of the concept set and the strength of the final concept.

The top-level activity map provided includes developing a strong concept, developing preliminary technical and cost models, developing target values for the system, identifying subsystems and their interfaces, and defining opportunities for the subsystems.

Two of the main messages in this chapter are that (i) the act of developing a strong concept is one of the most important things we do during product development, and (ii) developing that concept requires us to gain a rough understanding of what the major components or subsystems are and how

---

[13]An approach for evaluating a concept set is described more fully in Section 5.4.

they need to perform in order for the whole concept to be desirable. These two messages are essential, because they lay the ground work for the next stage of development, which is the subsystem engineering stage.

5.7   Exercises

Test Your Knowledge

T5-1 List five main elements of a high-quality system architecture.

T5-2 List the required content of the requirements, tests, and design at the end of the concept development stage.

T5-3 List three elements of an interface definition.

T5-4 List five fundamental steps in developing a strong concept.

T5-5 List four attributes that can be evaluated to assess the adequacy of a concept set.

Apply Your Understanding

A5-1 In your own words, define a strong concept.

A5-2 Evaluate some existing products relative to the performance measures in Exercise A4-6. Also, see if you can obtain market performance information about the evaluated products.

   a) Create a "House of Quality" (see Quality Function Deployment (11.51)) with the market requirements from Exercise A4-5 and the evaluation criteria from Exercise A4-6. Place the market performance information and the evaluations relative to the criteria in the House of Quality for the products you have evaluated.

   b) Create a scatterplot of market performance vs. performance measure value for each of the linkages shown in the House of Quality. Based on the scatterplots, propose ideal and marginal values for the criteria.

A5-3 Imagine you have been assigned to develop a new system for sharpening kitchen knives. The system should be usable by a homeowner.

   a) Find six existing sharpening systems by internet search.

   b) Find two existing sharpening systems by the Delphi method.

A5-4 Review the classification tree for the human-powered drill, found in Figure 5.14. Add a new branch and three concepts somewhere on the tree.

A5-5 Generate a candidate solution set for the problem of storing a snowboard or a surfboard in an apartment.

   a) Develop a concept classification tree for the concepts in the solution set.

   b) Assess the quality, novelty, variety, and quantity of your solution set.

   c) Do an intuition-based evaluation of the concepts. Are there any that you believe are promising to undergo further development? Why?

   d) Decompose the problem of storing a snowboard or surfboard in an apartment into a set of subproblems.

      a) List the subproblems.

      b) For each subproblem, identify four possible solutions.

      c) By combining one solution for each subproblem, create five concepts for storing a snowboard or surfboard. Describe each concept in words and sketches.

A5-6 Using the recombination table from page 272 in the Development Reference, generate and sketch 4 new concepts for a bicycle-like human transportation device.

A5-7 Thoughtfully prepare a screening matrix to evaluate a small set of promising concepts.

    a) How did the matrix influence your choice regarding which concepts to continue developing?

    b) What insights about your concepts did the matrix help you see, if any?

    c) In your own words, describe the risks you feel are associated with using a screening matrix?

A5-8 Write a brief stage report for Concept Development that can be used during the stage approval process for a project you're working on. Structure your report so that it answers these two fundamental questions: What is the selected concept? And how have you proven that your concept is best and that it meets the market requirements? As a way of supporting the report's claims attach and refer to product development artifacts that the team has produced.

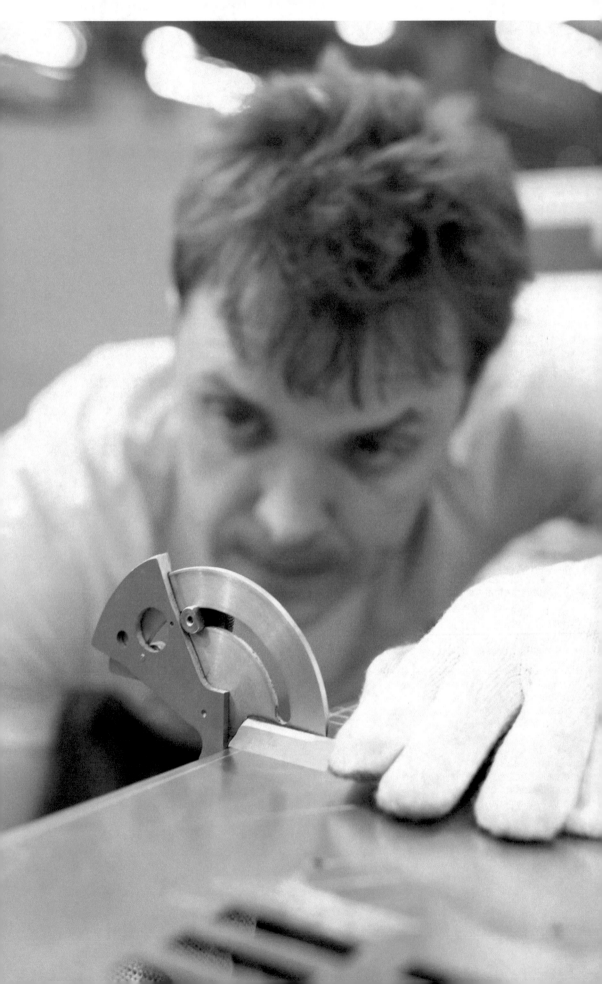

6

Subsystem Engineering

6.1 Design Evolution During Subsystem Engineering

The third stage of product development is subsystem engineering. In this stage, the concept's subsystems and components are fully detailed.

At the end of the subsystem engineering stage, the design is basically complete. Although it will be refined in later stages, it contains all the information necessary to create the system. The requirements and tests have also been updated to reflect analysis and testing of the engineered subsystems.

During the subsystem engineering stage, the system and its subsystems are being engineered – even though we simply call it subsystem engineering. To be clear about how this and the next stage of development are separated, we consider the end of subsystem engineering to be having tested and approved subsystems, while we consider the end of the next stage of development to be having a tested and approved system.

Table 6.1 provides a summary of the elements of this stage. Each is now described more fully.

Purpose

The purpose of subsystem engineering is to develop high-quality fully engineered subsystems and components that have been independently demonstrated to be desirable. Design for other characteristics such as manufacturability and ergonomics should have been accomplished. At the end of this stage the entire system has been designed, although a full system prototype has not necessarily been made or tested yet.

Development Outcomes

Transferable artifacts containing the following information for a desirable design must be approved at the end of subsystem engineering.

Requirements

There are two sets of information that must be added to the requirements by the end of subsystem engineering.

1. Predicted and measured performance values for subsystems.

   An important part of developing an engineered product is predicting how the product will perform, so that design choices can be made to achieve the

Table 6.1: Summary of the subsystem engineering stage.

Subsystem Engineering: Create high-quality engineered subsystems that have been demonstrated to be desirable. Design for other characteristics such as manufacturability and ergonomics should have been accomplished. At the end of this stage, the entire system has been designed, although the integration of subsystems has not yet been demonstrated.

	Required information	Typical artifacts	Checking criteria	Approval criteria
Requirements	Predicted and measured values for subsystem performance measures.	Section E of the requirements matrix for each subsystem	All predicted values are present, even if the value is N/A? All measured values present?	The subsystems meet or exceed the target values of the subsystem performance measures. If a few of the targets are not met, performance is at least in the acceptable range.
	Predicted values for system performance measures	Section E of the system requirements matrix	Predicted values for all performance measures? Consistent with subsystem measured values?	The predicted values meet or exceed the target values for the system performance measures. If a few of the targets are not met, performance is at least in the acceptable range.
Tests	Test data demonstrating subsystem desirability (measured values for subsystem performance measures).	Reports on methods and results demonstrating the desirability of the subsystem designs. Plots showing variation of subsystem performance with changes in design parameters.	Are the test methods complete and correct? Is there enough detail for a third party to repeat the tests?	The tests have been carried out with sufficient quality and transferability to provide strong evidence for the measured and predicted values.
Models	Engineering models used to choose design parameters and predict values for subsystem performance measures.	Software source code with run results. Input files for commercial software with run results. Excel spreadsheets. MathCAD worksheets. Hand solutions.	Are the models reported in enough detail to allow a third party to use them?	Not approved directly; used with tests.
Prototypes	Prototypes used to help choose values of design parameters. Prototypes used to measure measured values of subsystem performance measures.	Experimental testbeds used to select design parameter values. Fully functional subsystems for testing measured values.	Are the prototypes appropriate for the intended use? Is the fidelity and workmanship appropriate?	Not approved directly; used with tests.
Design	Geometric, material, and other appropriate definition of the design for each subsystem.	Engineering drawings of all custom-designed parts. Specifications (and possibly ordering information) for all purchased parts. Assembly drawings for the system and all subsystems. Schematic diagrams. Piping and/or wiring diagrams. Block diagrams. PC board layout files. Flowcharts and source code for any software that is part of the design.	Does the design package meet the design intent of the team? Are all relevant standards met? Are all the necessary components included? Is the design package sufficient to allow a third party to correctly make the product and test its compliance with specifications? Are all elements under version control?	The design is sufficiently transferable to allow the creation of complete subsystems and their integration into a complete system by a third party.
	Complete bill of materials	Spreadsheet or database table that lists all known components	Is it complete? Does it have all necessary information? Is it clear and unambiguous?	The bill of materials is appropriate
	Useful tools:	Bill of materials, CAD modeling, checking drawings, design for assembly, design for manufacturing, design of experiments, design structure matrix, dimensional analysis, engineering drawings, ergonomics, experimentation, failure modes and effects analysis, fault tree analysis, finite element modeling, prototyping, sensitivity analysis, uncertainty analysis.		
	Common pitfalls:	Avoiding analysis; reinventing analysis; never doing analysis; poor experimental procedure; focusing on the prototype, rather than the design.		

desired performance. This is often done using models. While not every performance measure needs to have predicted values, the most critical values should be predicted. During its evolution, the design changes to ensure that the predicted values meet or exceed the target values. The most current predicted values are captured in a design artifact (e.g., a requirements matrix).

As the subsystem engineering stage nears completion and the subsystem design is complete, prototypes of the subsystems are created and tested. The values of the performance measures that are obtained in testing are recorded as measured values. Where the measured values are inferior to the target values, the subsystem design is adjusted to improve the measured performance. The measured values are also captured in a design artifact, such as in Section E of a requirements matrix.

Some measures will be unable to be predicted; others will be unable to be measured. When the team decides not to measure or predict a value, this should be recorded in the requirements matrix (for example, the value could be recorded as "N/A"). Making such an entry shows that the lack of a value is intentional, rather than an oversight.

At the end of subsystem engineering, predicted and/or measured values for all subsystem performance measures should be entered in the subsystem performance matrices.

2. Predicted performance values for the system.

During subsystem engineering, the subsystems have not yet been integrated into the system. Therefore, there are unlikely to be measured performance values for the system. However, based on the decomposition work done in the concept development stage, there should be models that relate subsystem performance to system performance. Using the measured subsystem performance values, system performance

values should be predicted and compared with the system target values. The predicted values should be captured in a design artifact as well. If requirements matrices are being used, the most recent predictions should be recorded in the system performance matrix.

Tests

During subsystem engineering, models are tested to predict performance and prototypes are tested to measure performance. Information on this testing must be captured in a transferable format.

The methods used to predict product performance are an important part of product development (specific methods are explained more fully in Section 6.5). They too will evolve throughout the product development process. As they are used to test the current design, the results should be archived or if using a requirements matrix, included in the predicted values for the system or subsystem.

Prediction methods should be explained with sufficient detail that a skilled engineer who is not part of the development team could accurately apply the method based on the current design. Prediction methods should include the names of program files and/or data files used, along with any instructions necessary to use these files.

Not all performance measures need performance prediction methods. However, key performance measures should generally be supported by prediction methods. It is often useful to have at least two different prediction methods for critical measures. This increases confidence that the prediction is correct.

When predicting performance, it is generally a good idea to vary the parameters of the model to produce curves or surfaces, rather than single-point solutions. With curves or surfaces, the robustness of the prediction with respect to the model parameters can be estimated.

As prototypes are created, they should be tested to determine the measured performance on each of the performance measures. The procedures used in testing the prototypes should be recorded.

Test procedures should include test equipment, descriptions of how the testing is performed, the data that should be recorded, and the analysis methods for the data. There should be enough detail that a person not on the team could repeat the testing work and obtain the same results.

Well-documented tests of models and prototypes, with results that are consistent with the target values, provide strong evidence that the design is of high quality.

Models

An essential aspect of engineering design is to use mathematical models to predict the behavior of real systems. In subsystem engineering we choose values for critical design parameters by modeling the subsystems and then adjusting the design parameters until the model predictions match the target values for the subsystems. More detail on engineering modeling is given in Section 6.5.

Engineering models range from handwritten equations to computer-based models involving millions of lines of code. Regardless of their complexity, models should be transferably captured such that the modeling work could be repeated by a third party.

For models made using commercial software, this means that a set of input files should be kept and their use documented. For models that are created by the team, files that are part of the model, such as source code and data files, should be added to the product information. Hand calculations should be added to the product information, rather than kept on scratch paper or solely in and individual's notebook.

Like all product information, the models should be placed under revision control.

Prototypes

Many different kinds of prototypes are used during subsystem engineering. In this section we focus on two specific types: prototypes used to understand phenomena related to the subsystem and prototypes used to measure the performance of the subsystem.

As subsytems are engineered, sometimes not enough is known about the behavior of the subsystem to immediately create a single desired design. In such cases it can be helpful to create a *testbed*, which is a prototype that is designed to have the ability to quickly change the values of important design parameters. The testbed is then tested with various design parameters to determine the optimal design for the final product.

For example, when developing a Baja SAE vehicle, a team wondered about the effect of the location of the center of gravity (CG) on hill-climbing and steering performance. They created a testbed car that was too heavy to be competitive, but could readily have its center of gravity location changed. After exploring the effect of CG location with the testbed, they developed a competition car that was lighter but had much less adjustability.

After the subsystems are fully engineered, a final prototype of the subsystem is created using materials and processes as close to the final product intent as possible. This prototype is then tested to determine measured values of the subsystem performance measures.

Information about these prototypes should be placed under revision control and added to the product information.

Design

The following two main elements of the design are created or updated during subsystem engineering.

1. Definition of the system and subsystems

   By the end of the subsystem engineering stage, the design contains all the information necessary to make or buy the components and assemble them into the subsystems and integrate those subsystems into the system. The design will often contain many of the items listed below. All items included in the design should follow relevant professional standards. To greatly reduce confusion within the product development team, all artifacts that are part of the design should be placed under revision control, with a version number listed on the artifact and a history of changes maintained.

Assembly drawings: The overall system is represented with exploded and unexploded assembly drawings of the final product. These drawings should include critical assembly dimensions, overall footprint or size, assembly tolerances, notes, and a bill of materials defining each part in the assemblies. A numbered balloon points to each part of the assembly. There will likely also be assembly drawings for subsystems or other subassemblies.

Detailed part drawings: Parts should be described in sufficient detail to purchase or make them. Any custom-designed parts need engineering drawings. Drawings should include complete dimensions, tolerances, material types, and appropriate notes.

Schematic diagrams: When products contain electrical, hydraulic, or pneumatic systems, the design of such systems should be documented. Standard practices for schematic diagrams should be used, including standard symbols for components.

Wiring/piping diagrams: For products with electrical, hydraulic, or pneumatic systems, there may need to be a wiring or piping diagram in addition to schematic diagrams. Schematic diagrams are aimed at describing the function of the system; wiring and piping diagrams are aimed at describing the physical layout of the system. It will require judgment as to whether both types of diagrams should be provided. Schematics should always be provided. Wiring and piping diagrams may also be added if necessary.

Board layout: Products will often contain custom circuit boards. For such products, a copy of the board layout should be provided. The board layout may include a graphical image of the board. It may also include a variety of files used to produce the board, such as Gerber files and drill files. The board layout should be provided with sufficient detail that the board can be reproduced without reverse engineering.

Block diagrams: For electrical systems, it is common to create block diagrams, which show the design at a higher level of abstraction than the schematic diagrams. Block diagrams may also be useful in fluid control systems.

Logic diagrams: Products often contain software as part of a control system. When software is included, logic diagrams of the software should be part of the drawing package. This may include diagrams such as flowcharts, ladder diagrams, pseudocode, and block diagrams. The objective is to document the logic that is implemented in the software.

Source code: Where the design includes a software component, the source code should be provided in the form of computer files (not printed).

CAD models or IGES files: For complex shapes, it may be preferable to define the shape with a solid or surface model, rather than a drawing. The model files used to define the shape are an important part of the design, and should be referenced in notes on the drawing and in the bill of materials.

2. Bill of materials

A bill of materials (BOM) is a table showing all parts in the design in a hierarchical structure with top assembly, subassemblies, and parts. The BOM also has information regarding each part such as a part number, a drawing number, and a source for the part.

The BOM must be complete enough to allow a third party to aquire all purchased parts and the raw materials for all custom-made parts.

The BOM should be placed under revision control.

More information on the BOM can be found in the Development Reference under Bill of Materials (11.3).

Approval

The requirements, tests, and design are checked by the product development team to ensure that the product development artifacts accurately and transferably capture the design intent of the team. As the design artifacts will be used to transfer the design, it is essential that the artifacts be carefully reviewed, rather than just assuming they are correct.

Engineering models used as part of the tests create predicted values of performance. Ideally, these predictions should meet or exceed the performance targets. At a minimum, they should lie in the acceptable range.

Subsystem prototypes are tested to determine measured values of performance. Ideally, the measured values should meet or exceed the performance targets. At a minimum, they should lie in the acceptable range.

Every performance measure for the subsystem should have a predicted value and/or a measured value. The strongest proof of desirability is to have measured values, but in some cases only predictions can be made at this point. Where measured values are not available, predictions of performance should generally be made using at least two different prediction methods. Any performance measures that have neither a predicted nor measured values are cause for significant concern and are likely to prevent approval.

It may be difficult for the market representative to validate the design, because the subsystems are not yet fully integrated. However, it is a good practice to do as much as possible to create prototypes that are useful for validation. In particular, if critical subsystems can be evaluated by the market representative, the risk of developing an undesirable product is greatly reduced. For approval, the validation testing results should show that the market representative likes the design, with any areas of concern related to portions of the design not yet integrated in the validation prototypes.

Useful Tools

For subsystem analysis: Design for assembly, design for manufacturing, design of experiments, design structure matrix, dimensional analysis, ergonomics, failure modes and effects analysis (FMEA), fault tree analysis, finite element modeling, CAD modeling, prototyping, uncertainty analysis.

For subsystem testing: Prototyping, design of experiments, experimentation, prototyping, sensitivity analysis, uncertainty analysis.

For subsystem design: Bill of materials, CAD modeling, checking drawings, design structure matrix, engineering drawings, sketching, storyboards, robust design.

Common Pitfalls

- Avoiding analysis: It is virtually certain that assumptions and simplifications will need to be made in order to analyze the problem. Because this can be difficult, it is sometimes tempting to avoid analysis and do everything by experimentation. Even simple analysis can often lead to significant learning, such as identifying the important variables in the system.

- Reinventing the analysis: Engineering analyses of typical systems are often available in handbooks such as *Roark's Formulas* (Young et al., 2012) or *Marks' Standard Handbook for Mechanical Engineers* (Avallone et al., 2007). Machine design textbooks such as Norton (2006) or Budynas et al. (2011) can also be helpful. Be sure to explore available analyses before jumping into your own.

- Poor experimental procedure: Subsystems should be tested under a variety of conditions, not just under optimal conditions. It is best to identify the limits of satisfactory performance, not just verify that satisfactory performance exists under ideal conditions. Statistical Experimental Design techniques should be used to obtain as much information as possible from a limited number of experimental tests.

- Never-ending analysis: No matter how good the analysis model is, it can always be improved. However, the improvement is sometimes not necessary. When the analysis and its answers are good enough to support the necessary design decision, it's time to move on to the next critical issue.

6.2   Example of Subsystem Engineering

Let's again revisit the human-powered water well drill (presented more fully in AppendixB).

During subsystem engineering, all three components of the design (requirements, tests, and design) for the human-powered drill underwent evolution.

The requirements evolved and were captured by adding measured and predicted values to the subsystem requirements matrices. For brevity, we show only the requirements matrix for the wheel subsystem (see Figure 6.1). The measured and predicted values were obtained by applying evolved tests to models and prototypes developed from the design.

The team used two engineering models to test performance for the drill. These models were used repeatedly as the design of the drill changed, with the final predicted values being consistent with the target values.

The first test method was a model of the wheel RPM as a function of the torque required to cut through the earth. By examining existing research, the team discovered that a typical human could provide about 0.2 hp for a period of 2.5 hours[1]. Given this power output, the team considered the total power available as function of the number of operators from two to eight. The power to the drill is given by the product of the torque and the angular velocity. For four operators, the torque varied from 150 lb-ft at 30 rpm to 2,500 lb-ft at 2.5 rpm. These torque numbers could then be used to design the remainder of the subsystem.

Another example of a model used to test the drill is a stress analysis on the structure. Assuming that the drill string weighs 1725 lb and including the possibility of the drill string catching on the borehole side as it is being lifted, a total design load of 4500 lb was chosen for the string. With this load and the specified geometry and material for the structure, a safety factor of 1.5 for yielding was calculated.

The five subsystems on the drill were designed in an interdependent, rather than independent, fashion. Many of the key performance measures could only be determined in an assembled drill system, so multiple iterations of subsystem design, system integration, and system testing were performed during the subsystem engineering stage.

As a result of testing the models and prototypes, the design evolved throughout the stage. The design at the end of concept development included CAD models of the various subsystems. Engineering drawings were created for each of the parts identified, and assembly drawings were created for the subassemblies.

---

[1] http://www.ohio.edu/mechanical/programming/hpv/hpv.html

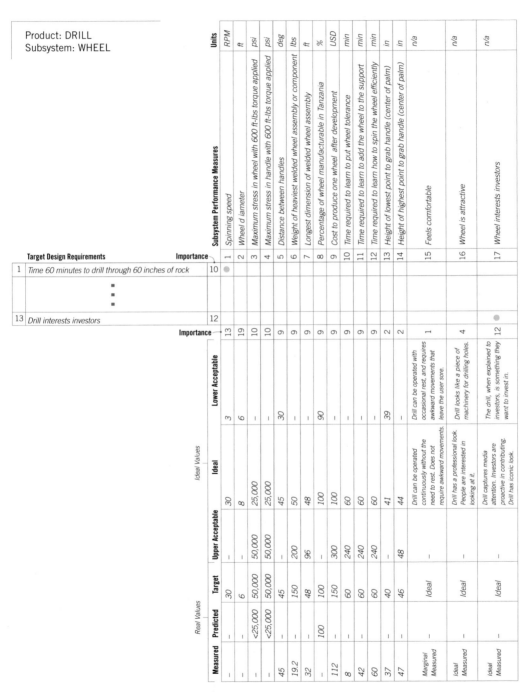

Figure 6.1: Subsystem requirements matrix with measured values for wheel subsystem. Because of space limitations, the market requirements and requirement–measure limitations are mostly omitted in this view.

By the end of the subsystem engineering stage the subsystems for the drill had evolved to the point shown in Figure 6.2. As can be seen, the subsystems are very similar to (but not exactly the same as) those at the end of the project (refer to Appendix B for images of the final design).

The design is captured in the form of engineering drawings. The team created 44 engineering drawings for this project, some of which had multiple sheets. Nearly all of these drawings were created and checked in the subsystem engineering stage of development. As an example, we have provided just four drawings; these drawings capture the design for the kelly bar assembly. These are shown in Figures 6.3 – 6.6. The complete set of engineering drawings organized by subsystem is shown in Figure 6.7.

By the time the stage of development was complete, the team had been able to predict or measure the performance of the subsystem as it related to each performance measure. In Table 6.2 we show the measured and predicted values for the structure subsystem. In all cases but one, the final values lay within the acceptable region. The sole exception is the parts count, where the upper acceptable limit is 100 and the actual number of parts is 107. In consultation with the sponsor, the team determined that the upper limit was somewhat elastic, and that an actual count of 107 would be acceptable.

* * * * *

As a note, the realities of an immovable project completion date required the team to reevaluate the scope during the early portions of subsystem engineering. Working closely with the client, and after considerable evaluation, the team and client decided to use a gasoline powered slurry pump in place of the human-powered pump being pursued by the team. While this was a difficult decision to make, it was clear that continuing development on the pump would jeopardize the successful completion of the other subsystems.

6.3 Top-Level Activity Map for Subsystem Engineering

Figure 6.8 shows the top-level activity map for the subsystem engineering stage of development. This activity map begins with the output of the concept development stage — approved product development artifacts including the concept, subsystem definitions, interface definitions, and target values for performance.

The initial activity shown in the map is to assign resources to subsystems that will allow the engineering of individual subsystems in parallel, at least to the point that the subsystem is ready for integration testing with some other subsystems. In some cases, subsystems will be approved only after successful integration testing with other subsystems. In others, the subsystem can be individually approved.

Compound Outcome 26 simply indicates that *all* subsystems are approved only after *every* subsystem is individually approved.

Importantly, the five arrows proceeding from Outcome 1 represent the engineering of each subsystem. The activity map for engineering a subsystem is too complex to have included in this figure, so that entire activity is detailed out in Figure 6.9, which shows a top-level activity map covering the independent engineering of a single subsystem. The map begins with identifying a list of all the components in the subsystem in a working bill of materials for the subsystem. The components are then classified as vital or mundane and custom or off-the-shelf. Each class of component has a different path; at the end of each path the component is integrated into the subsystem design. Once the subsystem design is complete, performance testing is used to demonstrate that the subsystem is ready for any needed integration tests.

This activity map shows one common relationship between subsystems, namely that engineering can proceed independently until integration tests are necessary. However, it is also possible that there will be interdependencies between other subsystems that may happen before

Figure 6.2: The drill subsystems as they existed at the end of subsystem engineering. a) drill string, b) hoist, c) wheel, d) wheel support, e) pump system.

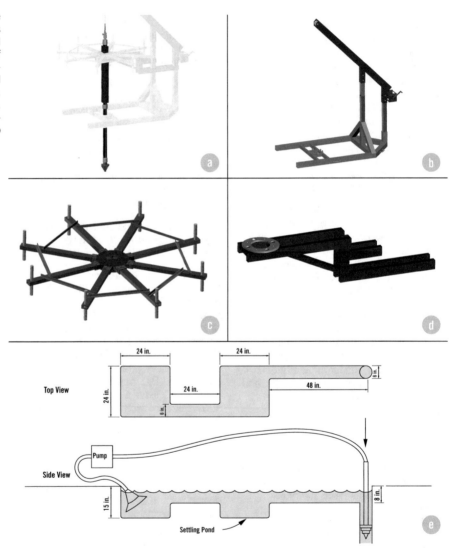

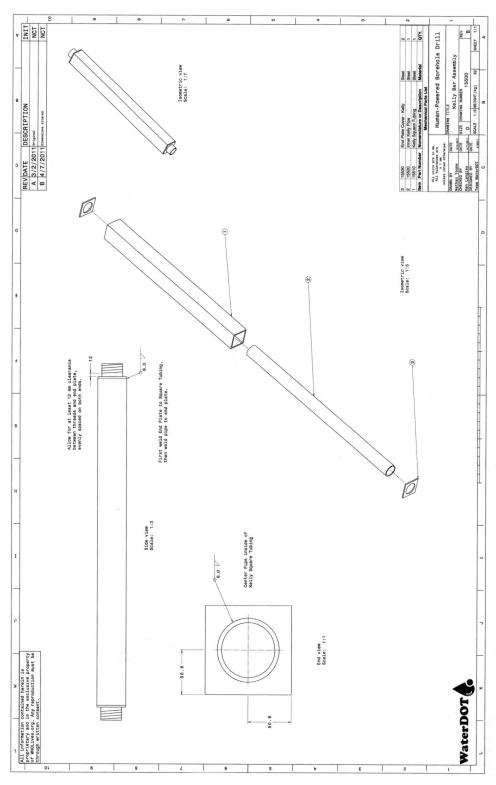

Figure 6.3: Engineering drawing of the kelly bar assembly.

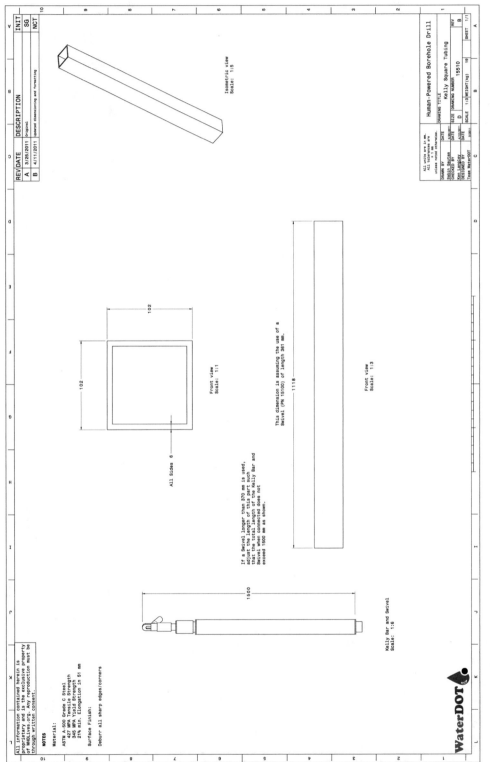

Figure 6.4: Engineering drawing of the square tubing for the kelly bar.

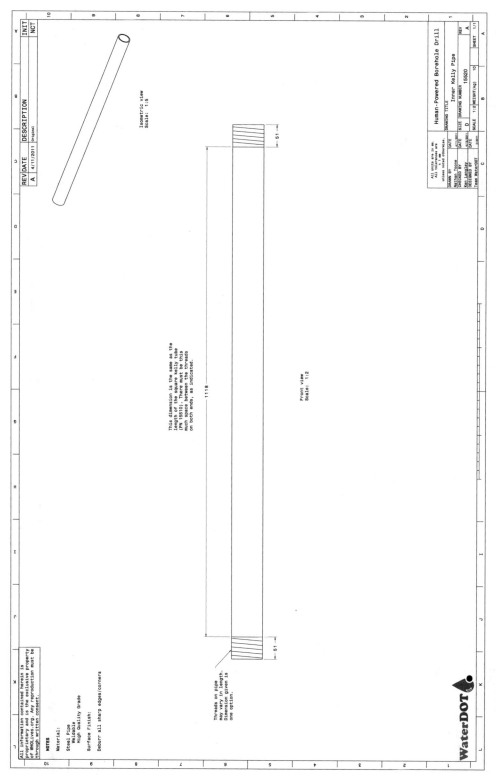

Figure 6.5: Engineering drawing of inner pipe for the kelly bar.

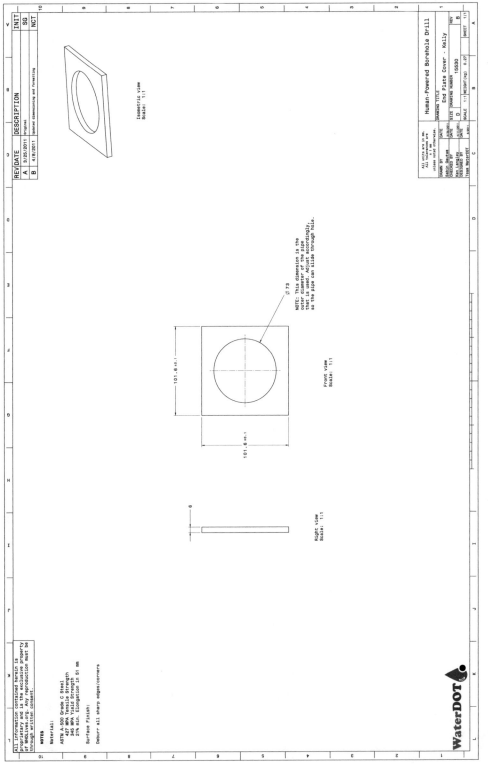

Figure 6.6: Engineering drawing of end plate for the kelly bar.

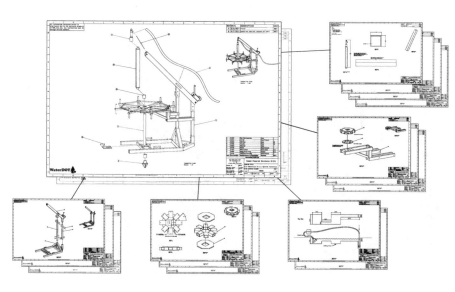

Figure 6.7: A representation of the 44 drawings created by the team, organized by subsystem.

Performance Measure	Ideal Value	Acceptable Limit	Target	Predicted	Measured
Maximum stress at depth (psi)	25,000	50,000	50,000	30,000	N/A
Load rating of support mechanism (lbf)	4500	2250	4500	N/A	3500
Maximum stress during changeover (psi)	25,000	50,000	50,000	1607	N/A
Downward force (lbf)	3000	500	200	N/A	1313
Rate of descent (ft/min)	4	1	4	N/A	3
Length of longest piece (in)	84	96	84	84	84
Percentage of structure made in Tanzania	100	90	100	97	N/A
Percentage of lifter made in Tanzania	100	0	100	0	0
Production cost (USD)	750	1000	750	842	N/A
Number of parts	30	100	30	107	107
Time to learn to set up (min)	60	240	60	N/A	18
Time to set up (min)	60	240	60	N/A	24

Table 6.2: Predicted and measured values for structure subsystem.

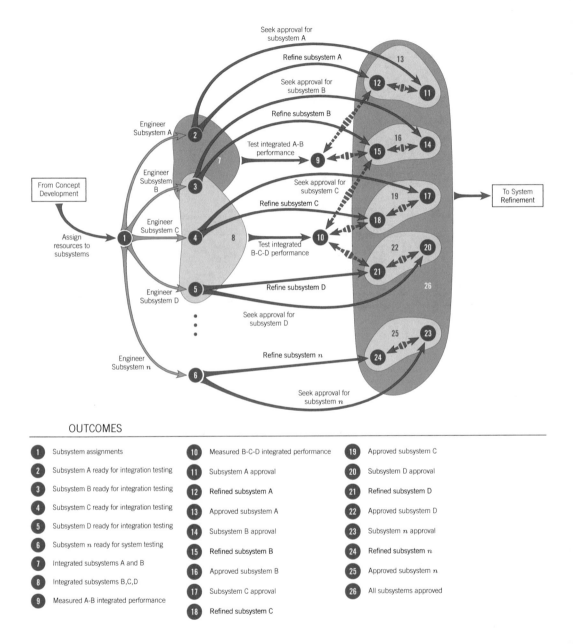

OUTCOMES

1 Subsystem assignments	**10** Measured B-C-D integrated performance	**19** Approved subsystem C
2 Subsystem A ready for integration testing	**11** Subsystem A approval	**20** Subsystem D approval
3 Subsystem B ready for integration testing	**12** Refined subsystem A	**21** Refined subsystem D
4 Subsystem C ready for integration testing	**13** Approved subsystem A	**22** Approved subsystem D
5 Subsystem D ready for integration testing	**14** Subsystem B approval	**23** Subsystem n approval
6 Subsystem n ready for system testing	**15** Refined subsystem B	**24** Refined subsystem n
7 Integrated subsystems A and B	**16** Approved subsystem B	**25** Approved subsystem n
8 Integrated subsystems B,C,D	**17** Subsystem C approval	**26** All subsystems approved
9 Measured A-B integrated performance	**18** Refined subsystem C	

Figure 6.8: Top-level activity map for the subsystem engineering stage. There is a great deal of complexity not shown in this map that covers the engineering of each individual subsystem. Please refer to Figure 6.9 for the detailed top-level map for engineering a single subsystem.

integration testing. Such interdependencies should be included in a specific activity map for the interdependent subsystems.

6.4 How to Design a Vital Custom Component

A common design activity is that of designing a vital custom component. The purpose for this activity is to develop a complete definition of the component, along with predicted and/or measured performance that demonstrates the component will work as intended.

An activity map for designing vital custom components is shown in Figure 6.9, between Outcomes 13 and 27. This section of the map covers preparing a component requirements matrix, identifying important engineering principles related to the component, identifying vital parameters of the component, developing analytical or experimental models that relate parameter values to performance, using the models to select parameter values, capturing the parameter values in an engineering drawing, and revising and checking the drawing.

Create the Component Requirements Matrix

A component requirements matrix has subsystem performance measures on the left and component design parameters on the top. The center of the matrix lists correlations between design parameters and performance measures, although the specifics of the mathematical relationships between the parameters and performance measures are not yet known at the start of component design.

One of the first activities in component design is to identify the subsystem performance measures that will be affected by the component. This should be relatively straightforward, because the component was chosen to create the appropriate performance.

A more challenging activity may be to identify the component parameters that

affect the performance. This will often require a literature search or discussion with experts who are familiar with the component technology and can identify important parameters. As these vital parameters are identified, the correlations between parameters and performance are also identified and recorded.

Identify Vital and Mundane Parameters

The vital parameters of the component are design parameters that strongly affect the subsystem performance. These will be listed at the top of the component requirements matrix. Values for the vital parameters will be carefully selected; determining these values will take up the majority of the time in subsystem engineering.

However, the vital parameters of the component are typically only 5-10% of the parameters for the component. Other design parameters have relatively minor influence on performance, and are called mundane parameters. Creating an initial CAD model, sketch, or drawing of the component can help identify all of the parameters necessary to be specified as part of the design. Values for the mundane parameters will generally be chosen conservatively, using good engineering judgment.

Determine Mathematical Relationships for Vital Parameters

Previously, we identified the vital component parameters that were correlated with subsystem performance. In this activity, we identify functional relationships that quantify the correlations. To do this we create analytical and experimental models as described in Section 6.5.

Predict the Performance and Choose Parameter Values

Once the parameter–performance relationships are identified, they are used to determine ideal values for the vital component parameters. This is done in an iterative process, often using optimization

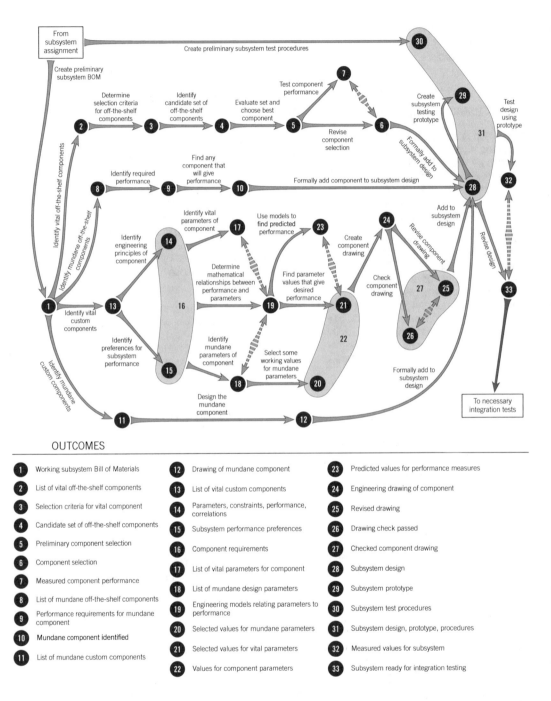

Figure 6.9: Top-level activity map for engineering a single subsystem. This map will be repeated for each subsystem. Included in this map are submaps dealing with mundane off-the-shelf components, vital off-the-shelf components, mundane custom components, and vital custom components. Note that a given subsystem may have zero, one, or more than one of any or all of these kinds of components.

techniques that adjust the component parameters until the predicted values for subsystem performance measures match the target values. This process is described more fully in Section 6.5.

Create the Component Design

Previously, we described creating an initial CAD model, drawing, or sketch of the component in order to identify all of the components design parameters. At this stage we create a formal engineering design, most often in the form of one or more engineering drawings, that captures all component design parameters, both mundane and vital. This design serves as the clear and complete definition necessary for the production system to create the component.

Check and Revise the Component Design

At the end of the activity map, the component's design is checked to ensure that it and the evidence of its desirability are transferable. The design should be checked by a member of the team other than the one who created the design. It's essential to look hard for mistakes in the design so they can be fixed at this time. The Development Reference entry on Drawing Checking (11.22) contains a checklist that is useful in checking the component design.

Once the design has been checked and approved by the drawing checker, the component design is added to the subsystem design.

6.5 How to Develop and Use Engineering Models

The goal of modeling is to understand the relationship between design parameters and performance measures.

A fundamental characteristic of developing engineered products is the use of predictive models to determine the expected performance of the product. Such models are used to quickly adjust the design to meet the required performance

characteristics. In most cases, the quickest and least expensive way to explore design parameters is with an engineering model.

In some form or another, modeling is used during most product development stages. Because many of the models are developed during subsystem engineering, we have included a deeper discussion of it in this chapter.

Common uses of modeling include the following:

- Identify the key physics involved in the problem to be solved

- Explore limits of the design space to help assess the novelty and variety of the candidate solutions

- Obtain quantitative estimates of the performance of candidate solutions to support the selection of a strong solution candidate

- Choose appropriate values of design parameters to ensure required performance is obtained

- Verify that the required performance is obtained.

What Is an Engineering Model?

Models are analytical approximations of the product, where the amount of approximation varies for different models. The degree to which the model matches the real product is the fidelity of the model. Models with high fidelity are a better approximation of the product than models with low fidelity. For example, Figure 6.10 shows various models of the mass distribution of a horse with different levels of fidelity. The model on the left (the lowest level of fidelity) would suffice if we were only concerned about the orbital characteristics of a horse, but if we wanted to examine stresses in horse's legs, we would need something more like the model on the right. To that end, it is important for the development team to match the fidelity of the model to its expected use.

Figure 6.10: Models of the mass distribution of a horse with various fidelities. The model on the left would serve if only the total mass were important and would allow rapid calculations. Models with higher fidelity would be required if the distribution of the mass were important for the question to be answered. Higher-fidelity models would require higher computational cost.

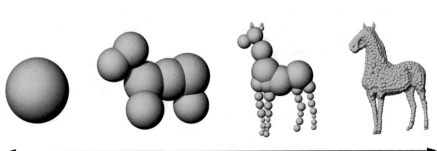

Low Fidelity Model
Low Computational Cost

High Fidelity Model
High Computational Cost

For the purpose of this discussion, an *engineering model* is a tool that can be used to express the performance measures of the product as a function of the values of the design parameters.

In its most general form, we can consider an engineering model to be based on an equation (or system of equations):

$$F(PM, DP) = 0 \qquad (6.1)$$

where PM is a set of performance measures, and DP is a set of design parameters. This form of the equation is not generally useful for design, so we desire to solve the general equation for the performance measures:

$$PM = G(DP) \qquad (6.2)$$

We can classify engineering models by considering the characteristics of the general equation and the solution. When symbolic solutions are available, we can use them directly. If there are no symbolic solutions, we must use a number of solutions to develop an empirical or statistical model.

Figure 6.11 shows the kinds of solutions that are likely to be available for various types of models. Note that model fidelity — which is a measure of how well a model represents reality — is generally considered to increase as we move from left to right on the figure. When we choose a particular kind of model, we are often committing to a specific fidelity, time required, and set of limitations. We are also choosing the type of insight we are likely to obtain from the model. Table 6.3 shows the general characteristics of each of the model types.

Planning for Engineering Models

In developing engineering models, it is wise to consider the following planning steps to ensure the best use of resources.

Define the purpose of the model: Why are we making a model? What performance of the product do we hope to predict? How will the model help us advance the design? Answering these questions helps ensure that the model is appropriately focused, and that the right kind of model is made.

Define the fidelity (level of approximation) of the model: Every model is an abstraction of the real world. Low-fidelity models have coarse approximations and consider limited physical phenomena. But their

	"Back of the Envelope" Model	Fundamental Engineering Model	Discretized Model	Physical Model
Symbolic solution available	●	●		
Numerical solution available, but not symbolic		●	●	
Experimental solution available, but not computational				●

Figure 6.11: Solution characteristics of model types.

Table 6.3: General characteristics of engineering models and their solutions.

Model and Solution Type	Fidelity	Time Required	Insight Gained	Limitations
Closed-form solution to simple model	Low to medium — often based on gross approximations	Low — can often be done "on the back of an envelope"	Important variables are identified, along with an estimate of their effects	Real products seldom closely match the simplifications assumed
Closed-form solution to engineering model	Medium to high	Low if found in handbook, medium to high if developed for your particular project	Detailed prediction of effects of important variables	Sometimes fine details are missing Often applies only to very specific cases
Empirical model, based on testing of physical prototypes	Depends on prototype that is tested, and number of tests done	High	Statistical model of performance, effects of unanticipated phenomena that didn't show up in the pre-test model	Testing can be very expensive. Statistical models are often just approximations of the real behavior. Extrapolation beyond the area tested is risky.
Numerical solution to engineering model (e.g., Runge–Kutta method for solving ODE)	As good as the fidelity level of the equation to be solved (generally medium to high). May be dependent on characteristics of the solution technique (e.g., step size)	Low, once equation is developed.	Insight comes in developing, rather than solving, the model. Systematically varying design parameters can lead to an empirical estimate of performance	Sometimes fine details are missing. Often applies only to very specific cases. Extrapolation beyond the area of systematic exploration is risky.
Finite element methods (and similar techniques)	Various, depending on the features included and the size of the mesh	Moderate time, depending on the complexity of the mesh	Need systematic variation in order to get good predictive insight.	Often limited to linear behavior. Very dependent on model setup. Model setup can easily be poor. The full set of features may give spurious solutions. Fine meshes can lead to very long solutions times.

corresponding strength is that they are quick to develop and usually quick to evaluate. Higher-fidelity models capture more physical phenomena and use finer approximations, but generally take more time to develop and execute. Higher fidelity is not better for all applications. Choose the right fidelity for your purpose. It is often wise to think about planning for a series of models with increasing fidelity.

Define the experimental plan for the model: Once the model is developed, what do you plan to do with it? Generally the model will be run multiple times with different values of model parameters. Single answers from a model are not very useful. Much more useful is a series of runs that can be used to develop operating curves or surfaces. Engineering models can be used with designed experiments to develop approximate response surfaces. Make sure you plan to use the model effectively.

Define the schedule for the modeling activity: Like any design activity, modeling can be endless if it is not managed properly. Plan a schedule to limit the amount of time spent on the model to an appropriate value.

Principles of Engineering Modeling

Just as in physical prototypes, the quality of engineering models improves as successive iterations are made through modeling activities. The IDEO adage "fail often early to succeed sooner" applies to engineering models. Most of the improvement happens when you make the first model; the improvement continues with the second. Eventually the improvement tails off, as is shown schematically in Figure 6.12.

Models need to be verified before they can be believed. The existence of a model, even using a commercial analysis program, does not imply that the model is correct. As models are being developed, it is important to check and see if the results make sense.

For critical performance measures, it is a good practice to have at least two independent ways of predicting and verifying the performance obtained. This increases the confidence in the quality of the proposed solution.

General Kinds of Models

Figure 6.11 presents four general kinds of models. They are:

Simplified "back of the envelope" models: Based on fundamental scientific principles and gross approximations.

Fundamental engineering models: Based on fundamental scientific principles, but with more refined approximations.

Discretized models (FEM, Finite Difference): Based on models applied to discrete portions of the system, with system performance calculated from the collection of the discrete portions. Discretized models are often used when the geometry or boundary conditions are too complex to create a fundamental engineering model. The system is discretized, and appropriate numerical models for each of the discrete elements are created, with consistent boundary conditions between the elements. All of the discrete models are solved, which leads to a numerical solution for the entire system.

Physical models: Based on experimental results. These are used when it is unclear how to create the other model types of the system (often because of uncertainty in how physical principles apply).

Solutions to the Model

The availability of a solution is dependent upon the type of model. There are generally three different solution regimes.

1. The solution is available symbolically (back of the envelope and fundamental engineering models).

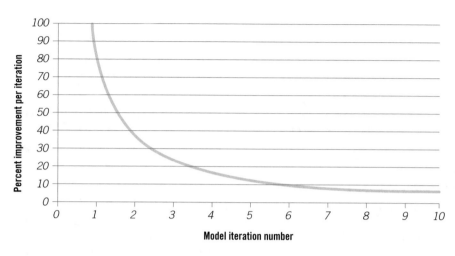

Figure 6.12: Schematic of how models improve with successive iterations. The biggest step in getting a model is the first iteration; the second iteration is also likely to have a large improvement. Eventually the improvements become relatively small.

2. The solution is available numerically, but not symbolically (some fundamental engineering models and discretized models).

3. Solution is available experimentally, but not computationally (physical models).

In planning for your models, be sure to account for the kind of solution you need and the kind of solution you'll get from the model.

Finding General Equations for the Model

One of the challenges in engineering modeling is determining the appropriate equations to use to model a real-world situation, as all general equations involve simplifying assumptions. Below are a few tips that might help as you develop models.

- Back of the envelope models: Look for fundamental laws that should apply (Newton's laws, conservation laws, etc.). It is often helpful to consider static or quasistatic conditions before developing dynamic models.

- Fundamental engineering models: Look in engineering textbooks and design handbooks, such as *Roark's Formulas for Stress and Strain* (Young et al., 2012). For higher-fidelity models, you may wish to do your own derivation, starting with related existing cases.

Using Model Solutions

Once a model has been developed and solved, the task is not yet done. We still need to use the solution to affect the design. Here we provide some guidelines for using each type of solution.

Symbolic solutions Look at the functional form of the solution. You can make great inferences from the functional form. For example, can you identify which parameter directly affect the solution? Can you see which ones inversely affect the solution? You can also create curves and surfaces of the solution function, and vary these plots by changing design parameters.

Numerical solutions These solutions cannot generally be used by direct inspection, so it will be necessary to generate enough solutions to draw inferences about the performance of the system. Ways that this is often done include the following:

- Use an optimization tool to optimize, as well as to explore the design space.

- Run the calculations multiple times and develop a statistical model that describes the performance (response surface methodology).

Physical models Run the experiments multiple times and develop a statistical model that describes the performance of the system. Design of experiments (DOE) methods are helpful here.

Summary of Engineering Modeling

There is a need to predict the performance of your product before it is built, so that a version of the product that meets the needs can be created.

There are different types of analytical models, depending on the type of fundamental information available about the system being modeled and the solvability of the developed equation.

As a team, you should develop a modeling plan to ensure that you are using your resources wisely.

6.6   How to Develop and Use Prototypes

The goal of prototyping is to create a physical representation of the product that can be used to help evolve the design.

As the design, requirements, and tests evolve through the product development process, it can be difficult for both the development team and the market representatives to understand and evaluate the current design. Manufactured products are inherently physical objects that we interact with using our physical senses. Designs are inherently intangible objects (although artifacts are tangible) that require us to process them before we can have a sensory experience with them.

To allow both the development team and the market representatives to have a direct sensory experience with the product, prototypes are made throughout the development process. Prototypes are physical representations of the design at the moment the prototype is created.

Like engineering models, prototypes have multiple purposes and can be of varying fidelity. Like models, a higher-fidelity prototype is not always better than a lower-fidelity prototype, given the time and cost necessary to produce a prototype.

Prototypes can represent the entire design (a *comprehensive* prototype) or only a small portion of the design (a *focused* prototype). Of course, they can also lie between these extremes.

Ulrich and Eppinger (2012, pp.294–297) note four purposes for prototypes, to which we add performance testing and validation, as shown in Table 6.4. Note that a single prototype can have multiple purposes. Also note that no prototype that requires a significant investment in time or money should be made without a clear understanding of its purpose(s).

The process of creating prototypes can consume a great deal of time and resources. To ensure that these are kept within appropriate bounds, a prototype plan should be created for each prototype that will consume significant resources. The plan should include the following (Ulrich and Eppinger, 2012):

- Define the purpose(s) of the prototype

- Establish the level of approximation of the prototype

- Establish the quantity to be built

- Establish the test plan for the prototype

- Establish the schedule for building and testing the prototype.

Care should be taken to make effective use of resources in prototyping. The most obvious means of prototyping may not be the most effective. For example, although it may seem that a hand-wired circuit board may be the quickest way to get a prototype board created, commercial services that produce a small quantity of

Purpose	Description
Learning	Understand more about how the design ideas will work in the physical world. These are often related to physical engineering models.
Communication	Share with others the current state of design. Allows people to obtain sensory input about the product.
Integration	See how well the parts of the design work together. The performance testing and validation prototypes at the end of the system refinement stage are examples of integration prototypes, but other partial integration prototypes can be effectively used in product development.
Milestone	Provide a concrete goal for the team to work toward. A milestone prototype can be a very effective way of getting the development team to finalize the design by a milestone deadline.
Performance testing	Provide a means for the development team to demonstrate that the performance measures (both subjective and objective) are being met.
Validation	Provide a means for market representatives to express their perception of how well the product meets market requirements.

Table 6.4: Possible purposes for prototypes.

professional-quality prototype boards from CAD data are quick and relatively inexpensive. In fact, when considering the generally higher reliability of the commercial board it is often faster and cheaper. Similarly, instead of machining a part from aluminum, a rapid prototyped plastic part may provide a similar function.

Particularly in the early stages of design, prototypes can be made of materials that will be significantly different from the final materials. See Prototyping (11.50) and Rapid Prototyping (11.52) in the Development Reference for more information on prototype creation.

Many product developers believe that effective use of prototypes to test ideas beyond the development team is a vital component of effective development processes. Getting prototypes in front of others is a great way to move your design forward.

6.7   Summary

During the subsystem engineering stage, the design progresses from a complete system architecture to finished designs for all the subsystems that have been demonstrated to meet their individual performance requirements. The subsystems have been designed to be compatible with one another to form the complete system, but the system has not yet been fully integrated or tested.

At the end of subsystem engineering, the design is generally complete. However, it will be subject to refinement during later stages of development.

During subsystem engineering, detailed designs of components and subsystems are created. The creation of the detailed designs allows modeling and testing to predict and measure performance.

Engineering models are used extensively during subsystem engineering to understand how the required performance can be achieved. A series of models with increasing fidelity is often used.

Prototypes are created to test the performance of subsystems and to obtain validation from market representatives. Careful planning for the prototypes ensures that the maximum benefit is reached.

6.8   Exercises

Test Your Knowledge

T6-1 List the required content of requirements, tests, and design at the end of subsystem engineering.

T6-2 List six activities used to detail a component.

T6-3 List five common uses of engineering models.

T6-4 List four steps for planning engineering models.

T6-5 List four general kinds of engineering models.

T6-6 List the elements of a prototyping plan that should be completed before starting on a prototype that will consume significant resources.

T6-7 List six purposes for prototypes that are identified in this chapter.

Apply Your Understanding

A6-1 Consider an engineering model that you have created or used in the past, possibly as part of a product development effort.

    a) What is the general kind of model you created or used?

    b) What were the intended and actual uses of the model (note: models are often used for multiple objectives)?

    c) How was the model useful in advancing the design?

    d) What were the main resource challenges associated with the model?

    e) What were the limitations of the model?

A6-2 Consider a product development project with which you are familiar (perhaps one on which you are currently working). For that project, identify a model that can be useful.

    a) What is the purpose of the model?

    b) Explain the fidelity of the model or models you intend to use to help the development effort.

    c) In broad terms, define the experimental plan for how you will use the model once it is developed.

    d) Estimate the time and resources required to support the effort of developing and using the model.

A6-3 Consider a product development project with which you are familiar (perhaps one on which you are currently working). For that project, identify a prototype that can be useful.

    a) What is the purpose of the prototype?

    b) Explain the level of approximation of the prototype you intend to use to help the development effort.

    c) Determine the number of prototypes to be built, and explain your decision.

    d) In broad terms, define the experimental plan for how you will use the prototype once it is developed.

    e) Estimate the time and resources required to support the effort of developing and using the prototype.

A6-4 Write a brief stage report for Subsystem Engineering that can be used during the stage approval process for a project you're working on. Structure your report so that it answers these two fundamental questions: What is the complete system design, including the subsystem designs and their integration? And how do you know that the subsystems meet their individual requirements and will be compatible with one another? As a way of supporting the report's claims attach and refer to product development artifacts that the team has produced.

## CHAPTER 7

Product Refinement

At the conclusion of subsystem engineering stage, the full system design is basically complete. All components and subsystems have been completely detailed, and the subsystems have been demonstrated to achieve their desired functions. From this point on, design evolution generally consists of refinements to the existing design, rather than the creation of whole new designs. The purpose of this chapter is to characterize the three main stages of refinement. They are:

- System refinement: The refinement aimed at improving weaknesses identified when the entire system is tested as a whole.

- Producibility refinement: The refinement aimed at improving product weaknesses observed during the start of production.

- Post-release refinement: The refinement aimed at improving product weaknesses found by the market place, after the product was released.

Although each of the refinement stages has a unique focus, the common elements of refinement apply to all. Therefore, all of these stages are discussed in this chapter.

7.1 System Refinement

The fourth stage of product development is system refinement. The purpose of this stage is to integrate the subsystems into a high-quality working system and test it. Importantly, the system is refined as necessary to resolve any issues that arise during testing of the integrated system.

A primary task of this stage of development is the interdependent testing of engineered subsystems. Because a comprehensive, fully integrated, analytical model of system performance is often prohibitive to create, performance testing of the integrated system is largely based on physical experiments. These experiments are essential because approved engineered subsystems may work well alone, but not function well together. This stage seeks to identify these dysfunctions and refine the subsystems and system so that the product meets the requirements.

To make the most of the system refinement stage, it is desirable to identify as many dysfunctions as possible. This generally means testing the system over a wide range of conditions, rather than just at the optimum conditions. Aggressive testing in system refinement helps prevent unpleasant surprises after it is released to the production system.

Table 7.1: Summary of the system refinement stage.

System Refinement: Integrate the subsystems into a demonstrated, high-quality working system. Refine the design as necessary to resolve any difficulties encountered during testing.

	Required information	Typical artifacts	Checking criteria	Approval criteria
Requirements	Measured system performance values	Section E of the system requirements matrix	Are all predicted values present, even if the value is N/A? Are all measured values present?	The system meets or exceeds the target values of the system performance measures. If a few of the targets are not met, performance is at least in the acceptable range.
	Market response to the product	Section F of the requirements matrix. Reports on customer response to the product, as measured by surveys, focus groups, interviews, or other direct interaction.	Has the market response been adequately assessed?	The market finds the product desirable.
Tests	Updated methods used to predict and measure the performance of the entire system (or product).	Reports on methods and results demonstrating the desirability of the product. Plots showing variation of system performance with changes in design parameters.	Are the test methods complete and correct? Is there enough detail for a third party to repeat the tests?	The tests have been carried out with sufficient quality and transferability to provide strong evidence for the measured and predicted values.
Models	Engineering models used to choose design parameters and predict values for system performance measures.	Model source code with run results. Input files for commercial software with run results. Excel spreadsheets. MathCAD worksheets. Hand solutions.	Are the models reported in enough detail to allow a third party to use them?	Not approved directly; used with tests.
Prototypes	Prototypes used to help choose values of design parameters. Prototypes used to determine measured values of system performance measures.	Experimental testbeds used to select design parameter values. Fully functional systems for testing measured values.	Are the prototypes appropriate for the intended use? Is the fidelity and workmanship appropriate?	Not approved directly; used with tests.
Design	Refined definition for the entire system.	Engineering drawings of all custom-designed parts. Specifications (and possibly ordering information) for all purchased parts. Subassembly and assembly drawings and instructions for all subsystems and the system. Schematic diagrams. Piping and/or wiring diagrams. Block diagrams. PC board layout files. Flowcharts and source code for any software that is part of the design.	Does the design package meet the design intent of the team? Are all relevant standards met? Are all the necessary components included? Is the design package sufficient to allow a third party to correctly make the product and test its compliance with specifications? Are all elements under revision control?	The design is sufficiently transferable to support the creation of the entire system by a third party.
	Complete bill of materials	Spreadsheet or database table that lists all known components	Is it complete? Does it have all necessary information? Is it clear and unambiguous?	The bill of materials is appropriate
Useful tools:		Bill of materials, CAD modeling, checking drawings, design for assembly, design for manufacturing, design of experiments, design structure matrix, dimensional analysis, engineering drawings, experimentation, failure modes and effects analysis, fault tree analysis, finite element modeling, prototyping, sensitivity analysis, uncertainty analysis.		
Common pitfalls:		Following poor experimental procedure; creating new and distracting performance measures; making poor trade-offs; creating new problems; substituting team judgment for the market.		

Table 7.1 provides a summary of the system refinement stage.

Purpose

The purpose of the system refinement stage is to integrate the subsystems into a demonstrated, high-quality working system. The design is refined as necessary to fix weaknesses that are identified when the integrated system is tested.

Development Outcomes

The following design information must be present at the end of the system refinement stage.

Requirements

There are two elements of the requirements that must be added by the end of system refinement.

1. Measured system performance values

The values for performance measures obtained in testing the integrated system must be recorded. As the design is refined during this stage, the updated measured values are also recorded.

As the design is refined, the predicted values for the system will change, and the predicted and measured values for some subsystems may change. Changes in the measured or predicted values should be reflected in revised product development artifacts, and revisions should be noted in accordance with the revision control procedures for these artifacts.

2. Market response to the product

At the end of the system refinement stage, a prototype will exist that is faithful to the final product design, although it was likely not created on the final production system. Because this prototype is so near the final product, it is ideal for determining the market response.

Throughout the product development process, the team has been evaluating the quality of the product through the performance measures. Now it is time to ask market representatives what they think about the *product*, rather than about the *design*. The results of this investigation form the market response.

The development team places the product in front of market representatives through interviews, surveys, focus groups, or other direct interaction. Market representatives are asked to evaluate the product, and the evaluation data is captured as the market response and stored in Section F of the requirements matrix.

It is vital that this evaluation be performed by market representatives, rather than the development team, although the development team facilitates the evaluation process.

Tests

Now that an integrated system is available, the entire system (rather than just independent subsystems) can be tested. The methods used to measure system performance should be recorded in transferable artifacts.

Results of the system testing are reported and recorded in the requirements matrix.

As the refined system is tested, changes in the tests may occur. Any such changes should be reflected in the product development artifacts, and revisions should be noted in accordance with the version control procedures for these artifacts.

Models

Models of system performance have generally been created in subsystem engineering. New models may be created during system refinement, and existing models may be changed to reflect a revised design. These models should be recorded in transferable artifacts.

Prototypes

The prototypes used during system refinement should reflect the latest revisions to the design and should be transferably captured in artifacts.

Design

There are two main elements of the refined design.

1. Refined definition for the entire system

As the subsystems are integrated, refinements to the subsystem definitions will be necessary. The refinements should be captured in design artifacts, and revisions should be noted in accordance with the revision control procedures for these artifacts.

The required design information should be the same as for subsystem engineering.

2. Complete bill of materials

The bill of materials should be updated (and tracked with revision control) to reflect the changes that occurred during system refinement.

Approval

To complete this stage, the design must be checked and found to be desirable and transferable.

Performance testing will need to have been performed on all system performance measures. The measured values ideally would meet or exceed the target values. In all cases, the measured values need to be within the acceptable limits.

Before this stage is complete, validation with an integrated system prototype needs to have been performed, and the market representative needs to have found the product desirable.

Common Pitfalls

- Poor experimental procedure. The system as a whole should be tested under a variety of conditions, not just under optimal conditions. We do this to try to identify the limits of operation. It's good to consider extreme practical conditions that could lead to failure.

- Creating new and distracting performance measures. It's easy to let new, unapproved performance measures distract us from the measures that have already been approved. Avoid this problem by carefully evaluating with the client the impact of adding new requirements and measures.

- Making poor trade-offs. Compromise is a natural part of product development. Trade-offs between components will have to be managed, as well as trade-offs between performance measures. To avoid making poor trade-offs, prioritize the requirements to match the market wants, and make decisions that reflect this prioritization.

- Creating new problems: Integrated products often have components that are highly interdependent. Each fix to a problem will possibly create new problems somewhere else in the system. Avoid this by examining the impact of a candidate change on each subsystem, adjusting as necessary.

- Substituting the team judgment for the market: Because the product has been thoroughly tested to see how well the performance measures match the target, it is tempting for the team to assume that all is well based only on the performance measures. However, the performance measures were created to substitute for the market desires, rather than being the market desires themselves. At the system refinement stage, it is vital to provide a sample of the product to the market representative(s) and obtain their independent assessment of the desirability of the product. The market judgment, not the team judgment, will ultimately determine the success of the product.

Top-Level Activity Map for System Refinement

Figure 7.1 shows a top-level activity map for the system refinement stage. There are

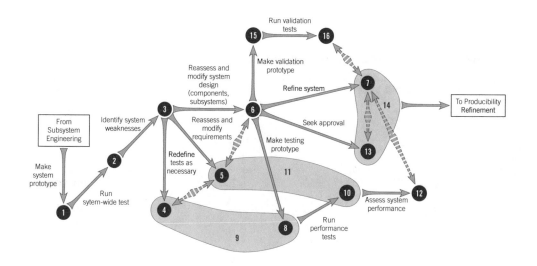

Figure 7.1: Top-level activity map for system refinement.

OUTCOMES

1 System prototype	7 Refined system design	12 System performance assessment
2 Measured system performance	8 Testing prototype	13 System approval
3 List of weaknesses	9 Testing essentials	14 Approved system
4 Improved tests and procedures	10 System measured performance	15 Validation prototype
5 Improved requirements	11 Assessment essentials	16 Market response
6 Improved system design		

three important things to observe from this map. First, the individual subsystems must be integrated for a system test, leading to Outcomes 1 and 2. Second, after any change is made to a subsystem, the entire system must be checked and tested as a whole (leading to Outcomes 9, 11, and 12) — and found to be desirable and transferable — before approval can be obtained. Third, as shown by the interdependent outcomes, iteration is often required until the desired system performance obtained.

Example

We now return to the human-powered water well drill to examine how this stage of development unfolded for the team. The initial parts of this stage were carried out completely in the CAD system, where the various subsystems were assembled together into a full system. Various small changes were carefully made to the subsystems and a full system emerged. The team evaluated the full system looking for weak points and refinements were made to improve them.

Table 7.2: Predicted and measured values for human-powered drill.

Performance measure	Ideal Value	Acceptable Limit	Target	Predicted	Measured
Maximum borehole depth (ft)	250	100	220	N/A	140
Time required to cut through 6 inches of rock (min)	45	60	60	N/A	480
Downward drilling force (lbf)	3000	500	200	1313	Not measured
Applied torque to pipe (lbf-ft)	400	200	300	1000	Not measured
Drill bit compatibility (%)	100	90	95	100	100
Water pressure down pipe (psi)	113	50	113	113	65
Water leakage through sides (%)	0	5	5	N/A	Not measured
Fraction of cuttings removed (%)	100	95	95	N/A	Not measured
Depth cut per 8 hours drilling (ft)	36	4	36	N/A	182
Number of people required to drill	3	12	4	4	4
Weight of heaviest subassembly (lbf)	50	400	200	332	
Longest dimension of biggest subassembly (in)	48	96	96	84	84
Fraction of drill producible in Tanzania (%)	100	85	90	N/A	66
Production cost (USD)	1000	5000	1500	1600	N/A
Time to learn to operate (hours)	4	20	3.5	N/A	3.5
Height of hand-operated parts (in)	3.5	Between 2 and 5	3.5	3.5	3.5
Drill feels comfortable	Can be operated continuously without rest	Can be operated with occasional rest	Can be operated continuously without rest	N/A	Can be operated with occasional rest
Drill is attractive	Has a professional look	Looks like appropriate machinery	Has a professional look	N/A	Has a professional look
Drill interests investors	Captures media and investor attention	Is interesting when explained	Captures media and investor attention	N/A	Captures media and investor attention

Figure 7.2: Acceptance test prototypes used during the system refinement stage of development.

of these refinements (and others) were motivated by weaknesses identified during system testing. The components and subsystem drawings were then updated and checked by the team.

A full set of drawings was created by the team (represented by Figure 6.7). They were released to a fabrication team and the parts were made and welded into subassemblies. The team faced very few problems getting the subsystems to work as intended. They owe this to the insight gained from building and testing various models and prototypes earlier in the development process. The full manufactured system, shown in Figure 7.2 as it existed during this stage of development, was tested by the team in various locations and under various conditions (as described in Appendix B). Multiple things were learned from performing those tests, including the actual performance achieved (see Table 7.2), and small refinements that could be made.

The market requirements were assessed by the client and by the users in Tanzania, who served as market representatives. Their evaluation of the market requirements is shown in Table 7.3.

7.2 Producibility Refinement

The fifth stage of product development is producibility refinement. During this stage, the design is refined as necessary to allow production at the rates and costs necessary for the final product.

Note that this is not the first time producibility has been considered. During previous stages, design for manufacturability and assembly were essential parts of the development process. However, as production moves from low-volume manufacturing to mass production, weaknesses in the product and process designs are almost certain to be discovered. As these weaknesses are found, the product and process designs are refined. As process design is outside the scope of this book, we focus on the refinements to the product design in this section.

For example, to keep the hose that transfers the cutting fluid (to the top of the kelly bar) from hitting the wheel as it spins, hooks were added to the design. Also the base of the hoist had open 4 inch by 4 inch square tubing. To avoid having the inside of the tubes exposed to moisture and therefore be more susceptible to corrosion, end plates were added to the design. Both

Table 7.3: Eval-
uation of mar-
ket requirements
by market repre-
sentatives.

Market requirement	Market representative evaluation
Drill reaches potable water beyond 100 ft	Excellent
Drill cuts through rock	Acceptable (slower than desired)
Drill uses existing drill bits	Excellent
Drill seals borehole sides to prevent cave in	Very good (no cave-ins observed)
Drill removes cuttings from the borehole	Excellent
Drill works at an efficient speed	Excellent
Drill uses only manual labor	Acceptable (gas-powered pump)
Drill is affordable	Very good (slightly high, but in good range)
Drill requires only simple training	Excellent
Drill is portable	Acceptable (heaviest piece is a bit heavy)
Drill is comfortable to operate	Very good
Drill is attractive	Excellent
Drill attracts investors	Excellent

Table 7.4 provides a summary of the elements of the producibility refinement development stage. The elements are explained more fully below.

Purpose

The purpose of this stage is to refine the design as necessary to allow a desirable product to be produced in the desired quality and quantity. A key feature of this stage is fixing producibility weaknesses that are identified during production ramp-up.

Development Outcomes

No new types of product information are created during this stage. However, refinements are made to the requirements, tests, models, prototypes, and design. These should all be captured transferably. Revision control is essential to ensure that the highest quality product can be made.

Producibility testing needs to have been performed that demonstrates the product can be produced at the required rate, quality, and cost.

Common Pitfalls

- Creating new problems. By this stage of development, many designs have been fine-tuned so that all subsystems are working well together. Even minor changes to one subsystem or component can negatively affect the whole product. When making changes at this stage of development, carefully examine each change's impact on every subsystem.

- Inadequate sample sizes. At this stage of the product development process, we often have access to a reasonable quantity of the product. As such, we should choose adequate sample sizes for testing and experimentation, so that their results are statistically representative of the body of products being produced.

Top-Level Activity Map for Producibility Refinement

Figure 7.3 shows a top-level activity map for the producibility refinement stage. There are two important things to note about this map. First, note that the stage is started when an initial production run of the product is made. As the product is made on the real production system, weaknesses in the design may show up and need to be resolved. Second, note that there is typically no market validation during producibility refinement, as the function of the product is not changed during this stage.

As in system refinement, iteration may be necessary as shown by the interdependent outcomes.

Table 7.4: Summary of the producibility refinement stage.

Producibility Refinement: Refine the design as necessary to allow a desirable product to be produced in the desired quality and quantity. Note that this stage is primarily about fixing producibility weaknesses that are identified during production ramp-up.				
	Required information	Typical artifacts	Checking criteria	Approval criteria
Requirements	Updated predicted and measured values of performance measures.	Part E of the system and subsystem requirements matrix.	Are all predicted values present, even if the value is N/A? Are all measured values present?	The system meets or exceeds the target values of the system performance measures. If a few of the targets are not met, performance is at least in the acceptable range.
Tests	Updated methods used to predict and measure the performance of the system. Methods used to demonstrate the producibility of the product.	Reports on methods and results demonstrating the desirability of the product. Reports of producibility challenges and their resolution.	Are the test methods complete and correct? Is there enough detail for a third party to repeat the tests?	The tests have been carried out with sufficient quality and transferability to provide strong evidence for the measured and predicted values.
Models	Engineering models used to analyze and adjust design characteristics related to producibility. Statistical analysis of producibility challenges (defects, low rate, high cost, etc.).	Model source code with run results. Input files for commercial software with run results. Excel spreadsheets. MathCAD worksheets. Hand solutions.	Are the models reported in enough detail to allow a third party to use them?	Not approved directly; used with tests.
Prototypes	Prototypes used to explore possible producibility solutions. Prototypes used to measure producibility of revised design.	Pilot-scale production runs with statistical analysis. Producibility studies on initial product runs and refined design.	Do the prototypes demonstrate the solutions to producibility problems?	Not approved directly; used with tests.
Design	Refined definition for the entire system.	Engineering drawings of all custom-designed parts. Specifications (and possibly ordering information) for all purchased parts. Subassembly and assembly drawings and instructions for all subsystems and the system. Schematic diagrams. Piping and/or wiring diagrams. Block diagrams. PC board layout files. Flowcharts and source code for any software that is part of the design.	Does the design package meet the design intent of the team? Are all relevant standards met? Are all the necessary components included? Is the design package sufficient to allow a third party to correctly make the product and test its compliance with specifications? Are all elements under revision control?	The design is sufficiently transferable to support production by a third party.
	Complete bill of materials	Spreadsheet or database table that lists all known components	Is it complete? Does it have all necessary information? Is it clear and unambiguous?	The bill of materials is appropriate
	Useful tools:	Design for assembly, design for manufacturing, design of experiments, experimentation, failure modes and effects analysis, fault tree analysis, plan-do-check-act, sensitivity analysis, six sigma, theory of inventive problem solving, troubleshooting, uncertainty analysis.		
	Common pitfalls:	Creating new problems; inadequate sample sizes; waiting until this stage to consider producibility.		

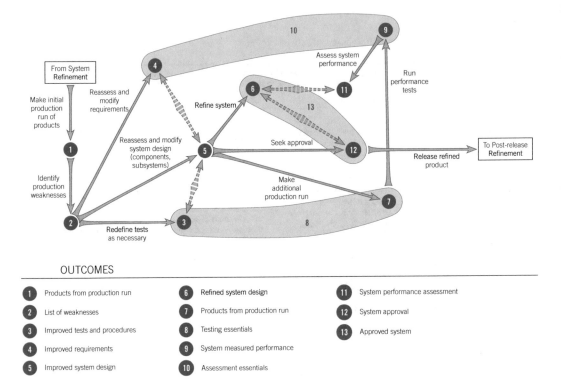

Figure 7.3: Top-level activity map for producibility refinement.

Example

To solidify the principles discussed in this section, we again consider the human-powered water well drill as an example. A different development team (based in Africa) worked on the producibility refinement and post-release refinement stages.

The design of the drill indeed underwent some small refinements due to manufacturing considerations, some of which were foreseen and some unforeseen. For the reason that nearly all of the first four stages of development were carried out in the USA, the development team worked with the stock materials that could be purchased easily in the USA. It was known and planned for that the manufacturing entity would convert the design to one that could be produced using

stock materials in Tanzania. During this stage of development, the design underwent these changes.

The drill was also refined in unexpected ways. For example, the original handles that were welded to the base of the hoist to assist in hoist transport, involved three pieces, four 45 degree cuts, two 90 degree cuts, and four welded joints per handle. This is shown in Figure 7.4a. The refined handle required only two 90 degree cuts and two welded joints. This was accomplished by adding two 90 degree bends to the piece that makes the handle, as shown in Figure 7.4b. The result is a much more manufacturable, but slightly less comfortable, handle. As manufactured cost is much more important than comfort of setting up the drill, this trade-off was deemed appropriate.

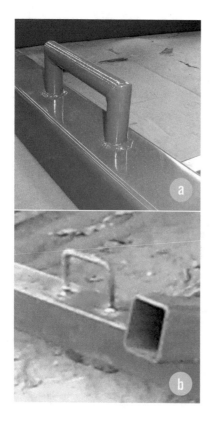

Figure 7.4: Small refinements made to the drill during the producibility refinement stage of development. (a) Original handle, (b) refined simplified handle.

Although the artifacts are not shown here, the essential outcome in terms of design information is the updated design — in this case the definition of the updated handle.

7.3 Post-release Refinement

The final stage of product development occurs after the mass-produced product has been released to the market. During this stage, the fully integrated, fully functioning, mass-produced product is refined until the owner of the design decides to end development. While this decision is often tied to a decision to retire the product, these decisions need not be related.

At the true end of development (or the end of the stage as seen in Figure 2.10), the final/updated product design is archived, sold, or removed from the context where design changes can be made. A wholehearted effort to accurately archive the design is more valuable than most people would believe, as the archived design is sometimes reopened for the purpose of starting a new design, or for legal purposes.

Unlike the other stages, this stage of development is not characterized by a specific evolution in the design. This stage may be thought of as a series of small stages, with one stage for each specific refinement in the product.

The need for refinement at this stage is often driven by one of the following:

1. The need to lower costs, enhance performance, or respond to changing market demands.

2. The occurrence of unforeseen problems with the product or its production.

3. The need to maintain a particular level of product performance.

Almost always there is a different budget for pre-release development (opportunity development through producibility refinement) and post-release refinement. Likewise, project milestones are often treated on an individual refinement basis when in post-release refinement.

Table 7.5 provides a summary of the elements of the post-release refinement stage.

Purpose

The purpose of this stage of development is to refine the design to improve the desirability of the mass-produced product. This includes items such as reducing cost, increasing functionality, and eliminating weaknesses that become apparent after the product has been offered on the market.

Table 7.5: Summary of the post-release refinement stage.

Post-release Refinement: Refine the design to improve the desirability of the mass-produced product. This includes items such as decreasing cost, increasing functionality, and eliminating weaknesses that become apparent after the product has been offered to the market.

	Required information	Typical artifacts	Checking criteria	Approval criteria
Requirements	Updated market and product requirements, including the additional information that was learned during this stage.	Updated subsystem and system requirements matrices.	Are all of the matrices complete and correct?	The changes in the market and product requirements capture the market's desires. The system meets or exceeds the target values of the system performance measures. If a few of the targets are not met, performance is at least in the acceptable range.
Tests	Updated methods used to predict and measure the performance of the system.	Reports on methods and results demonstrating the desirability of the product.	Are the test methods complete and correct? Is there enough detail for a third party to repeat the tests?	The tests have been carried out with sufficient quality and transferability to provide strong evidence for the measured and predicted values.
Models	Engineering models used to analyze and adjust design characteristics of the product. Statistical analysis of product weaknesses (defects, low rate, high cost, etc.)	Model source code with run results. Input files for commercial software with run results. Excel spreadsheets. MathCAD worksheets. Hand solutions.	Are the models reported in enough detail to allow a third party to use them?	Not approved directly; used with tests.
Prototypes	Prototypes used to explore possible improvements. Prototypes used to measure producibility of revised design.	Production runs with statistical analysis.	Do the prototypes demonstrate both the problem and the solutions?	Not approved directly; used with tests.
Design	Refined definition for the entire system.	Engineering drawings of all custom-designed parts. Specifications (and possibly ordering information) for all purchased parts. Subassembly and assembly drawings and instructions for all subsystems and the system. Schematic diagrams. Piping and/or wiring diagrams. Block diagrams. PC board layout files. Flowcharts and source code for any software that is part of the design.	Does the design package meet the design intent of the team? Are all relevant standards met? Are all the necessary components included? Is the design package sufficient to allow a third party to correctly make the product and test its compliance with specifications? Are all elements under version control?	The design is sufficiently transferable to support production by a third party.
	Complete bill of materials	Spreadsheet or database table that lists all known components	Is it complete? Does it have all necessary information? Is it clear and unambiguous?	The bill of materials is appropriate.
	Useful tools:	Design for assembly, design for manufacturing, design of experiments, experimentation, failure modes and effects analysis, fault tree analysis, plan-do-check-act, sensitivity analysis, six sigma, theory of inventive problem solving, troubleshooting, uncertainty analysis value engineering.		
	Common pitfalls:	Ignoring the market; creating new problems; inadequate sample sizes; failure to fully document everything.		

Design Outcomes

Post-release refinement often leads to updated market and product requirements. As the product matures, the desires of the market become more clear. This provides the opportunity to refine the requirements and to refine the product to better meet the requirements. As always, the requirements should be subject to revision control.

As needed, all other elements of the design are revised during post-release refinement. The revisions are captured in product development artifacts that are placed under revision control.

Useful Tools

Tools for continuous improvement: Design of experiments, experimentation, failure modes and effects analysis, fault tree analysis, plan-do-check-act (PDCA), sensitivity analysis, six sigma, troubleshooting, uncertainty analysis, value engineering.

Tools for the design process: CAD modeling, design for assembly, design for manufacturing, engineering change orders (ECOs), finite element modeling, prototyping, theory of inventive problem solving.

Common Pitfalls

- Ignoring the market: During this stage of development the team should consider happenings in the market to help drive decisions about product changes.

- Creating new problems: By this stage of development, many products have been fine-tuned so that all subsystems and components are working well together. Even minor changes to one subsystem can negatively affect the whole product. When making changes at this stage of development, carefully examine each change's impact on every subsystem.

- Inadequate sample sizes: At this stage of the product development process,

we often have access to an adequate quantity of the product. As such, we should choose adequate sample sizes for testing and experimentation so that their results are statistically representative of the body of products being produced.

- Failure to fully document everything: It can be tempting to just do the refinement, and get it into production, without doing all the documentation. Failure to document changes leads to ambiguity in the product design.

Top-Level Activity Map for Post-release Refinement

Figure 7.5 shows a top-level activity map for the post-release refinement stage. There are three important things to note about this map. First, note that the stage is started when weaknesses in the product appear after the product has been in the market. Second, note that there may or may not be significant market validation during post-release refinement, depending on whether the functionality from the user point of view has changed. In many cases, the objective of post-release refinement is to assure that the product continues to provide the performance identified during system refinement. Third, note that this map both begins and ends with post-release refinement; the map may be executed multiple times during the life of the product.

As in system refinement, iteration may be necessary as shown by the interdependent outcomes.

Example

The human-powered water well drill is currently in the post-release refinement stage of product development. This means the drill is in production and being used, and that from time to time improvements are made that evolve the design. The drill as it currently exists in this stage of development is shown in Figure 7.6.

As an example of the types of changes that typically occur at this stage of development,

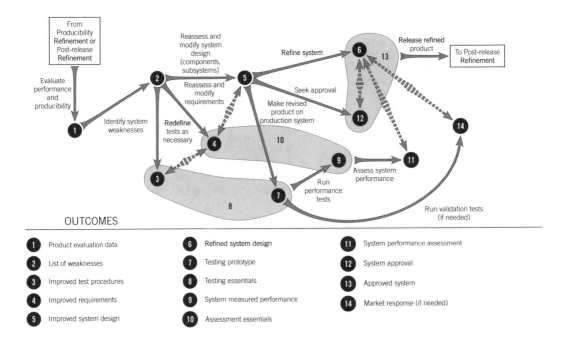

OUTCOMES

1 Product evaluation data
2 List of weaknesses
3 Improved test procedures
4 Improved requirements
5 Improved system design

6 Refined system design
7 Testing prototype
8 Testing essentials
9 System measured performance
10 Assessment essentials

11 System performance assessment
12 System approval
13 Approved system
14 Market response (if needed)

Figure 7.5: Top-level activity map for post-release refinement.

consider the lazy susan bearing that is the interface between the wheel and the wheel support. Figure 7.7 shows the bearing (has a silver color). The selected bearing can withstand sizable axial compressive loads, but it cannot withstand large tensile loads. While in use, it was observed that to get underneath the wheel to access the drill string, workers would hang on the wheel for support. Such movements created a tensile condition in the bearing on the opposite side of the wheel, often tearing the bearing apart. After considering a variety of solutions, the team chose to remove the bearing from the design and simply let the wheel spin directly on top of the wheel support. After evaluating the impact of this decision by considering how the market requirements were affected, the decision was approved. For existing drills, the bearings were removed when they failed in the field. New drills were produced without bearings.

7.4 How to Release a Design

Approved designs must be released to other parts of the organization that will use the design. The design release process is straightforward, as shown in Figure 7.8.

When a design has been approved, the approved revisions are then conveyed to all those who have need of it. This is done by ensuring that the revision numbers for all design artifacts are updated to reflect the release and that the change history that tracks the revision status of each component is also updated. Specific methods for updating and conveying the design will vary between organizations. However, some process should be followed to ensure that the latest approved revision is used throughout the organization. If needed, the Engineering Change Order (ECO) (11.24) and Revision Control (11.56) entries in the development reference can help you get started with revision control.

Figure 7.6: The human-powered water well drill, marketed as the Village Drill. This figure shows the drill as it exists during the post-release refinement stage of development.

Figure 7.7: The lazy susan bearing interface between the wheel and the wheel support for the human-powered water well drill.

Requirement and test artifacts may not have new revisions when a design is released. For example, test procedures may be standardized and may be used for a number of different products. Test results that were obtained on previous prototypes will not be revised. However, new tests may be performed on the released design which would result in new test reports and potentially revised requirements artifacts.

7.5 Summary

At the conclusion of subsystem engineering, the design is basically complete. However, it must be refined based on weaknesses that are observed (i) when the system is integrated and tested as a whole, (ii) when the system is produced in large quantities that expose previously unseen weaknesses, and (iii) after the product has been released to the market and the actual market response has been observed.

An essential part of refining a design is the release of a revised version of the design. When a design is released, its version and revision history are updated. Maintaining a complete revision history allows proper support for the design throughout the product lifecycle.

7.6 Exercises

Test Your Knowledge

T7-1 List the three main refinement stages.

T7-2 List two elements of design artifacts that are updated when a design is released.

T7-3 What is added to the requirements during the system refinement stage?

T7-4 What kinds of information are added (as opposed to updated) to the product development artifacts during producibility refinement and post-release refinement?

Apply Your Understanding

A7-1 In your own words, describe the differences between the three main refinement stages. Why do you think they are listed as three different stages, rather than being combined into a single stage?

A7-2 Why do you believe there are likely to be problems that show up in integrating the subsystems, even though each subsystem has been individually tested and approved?

Figure 7.8: A generic activity map for releasing an approved design.

OUTCOMES

1 Approved design   **3** Released design

2 Design with revision number

A7-3 Share an example with which you are familiar where changes to a design were found to be necessary during producibility refinement.

A7-4 Share an example of when post-release refinement is necessary (Hint: you may wish to consider product recalls).

A7-5 Find a design artifact (such as an engineering drawing) for a product with which you are familiar.

  a) Identify the revision and the revision history for the artifact.

  b) If these items are found in the artifact, what do you learn by reviewing the revision history?

  c) If these items are not found in the artifact, what is lost by their absence?

A7-6 List some reasons you believe having an appropriate revision history is helpful in providing long-term support for a product on the market.

A7-7 Write a brief stage report for System Refinement that can be used during the stage approval process for a project you're working on. Structure your report so that it answers these fundamental questions: What is the complete system design? How well does the system meet the requirements? And how have you validated system desirability to the market? As a way of supporting the reports claims attach and refer to product development artifacts that the team has produced.

A7-8 Write a brief stage report for Producibility Refinement that can be used during the stage approval process for a project you're working on. Structure your report so that it answers these fundamental questions: What is the refined definition of the system? And, how have you demonstrated that the refined system solves the producibility problems and still meets market requirements? As a way of supporting the reports claims attach and refer to product development artifacts that the team has produced.

A7-9 Write a brief stage report for Post-Release Refinement that can be used during the stage approval process for a project you're working on. Structure your report so that it answers these fundamental questions: What is the refined definition of the system? And, how have you demonstrated that the refined system solves the post-release problems and still meets market requirements? As a way of supporting the reports claims attach and refer to product development artifacts that the team has produced.

## CHAPTER 8

Customizing the Product Development Process

Ultimately, every product development team needs a specific, often custom, product development process that is catered to the needs of the client and skills of the team. Unfortunately, we cannot tell you in this book what specific product development process is needed for your specific project. This is simply because every project is unique and requires its own product development process and its own project plan.

We can, however, show you how to create, evaluate, and manage your own custom product development process and project plan. This chapter focuses on this.

To be clear, we'll define the custom product development process as the set of specific activities that need to be taken to achieve a specific set of design outcomes. Further we define the custom project plan as the custom product development process with resources (human, financial, and temporal) added to activities.

Every project should have at least some plan — even if that plan is not very detailed. Whatever level of detail is chosen for the project plan, the following sections will guide the team or team leader in establishing a good plan that is likely to lead to success.

During development, plans evolve. It is unrealistic to believe that a plan will first be set, then followed without adjustment. Interestingly, having an initial plan in place when unexpected problems or delays occur allows the team to know how the plan can change with minimal impact on the highest level project objectives.

8.1 Creating a Customized Project Plan

A customized project plan is (i) specific to the project at hand, (ii) aligned with the capabilities of the development team, and (iii) established according to the client's budgetary and time constraints.

The customized plan includes an activity map (set of design activities, outcomes, and their relationships) that will transition a design from its initial state of evolution to the client-specified final state of evolution. Ultimately, as the plan evolves over time, each of these activities and outcomes will be defined to such a degree that they can be assigned to an individual or subteam to complete. The activities will be defined with sufficient clarity that the assignee can complete it according to the budget and schedule for that specific activity.

C. A. Mattson, C. D. Sorensen, *Product Development*,
https://doi.org/10.1007/978-3-030-14899-7_8

The following steps can be used to create a customized product development process (steps 1–4) and transition that process into a custom plan (steps 5–6). Each one of the steps is described in more detail in the subsequent subsections.

Process for customizing the project plan:

1. Establish the project scope

2. Articulate specific end-of-stage outcomes

3. Identify specific intermediate design outcomes

4. Choose design activities that lead to the outcomes

5. Allocate resources to each design activity

6. Add time estimates and extract the project schedule.

Step 1: Establish the Project Scope

As a first step, the project scope should be established and stated succinctly and accurately. This must be done early in the project. The scope will not only guide the development of the customized project plan, and it will also guide the product development team throughout the entire project. The project scope can be captured in a project objective statement, which is a short statement about a project's top-level scope, schedule, and resources. A description of this and a method for creating one is found in the Development Reference, under Project Objective Statement (11.49).

An important part of this step is to bracket the development, which means to determine which portion of the product's evolution your team will be working on. This is done by (i) understanding the current state of the design, and (ii) understanding the required final state of the design. The current state and the required final state of the design exist somewhere on the evolutionary continuum defined by the stages of product development. Figure 8.1 illustrates this for the human-powered drill.

Step 2: Articulate Specific End-of-Stage Outcomes

Implied in bracketing the development is the completion of intermediate stages of development, each of which has a generic purpose associated with high-level design outcomes. For manufactured products that are expected to evolve through the six stages of development, Tables 4.1, 5.1, 6.1, 7.1, 7.4, and 7.5 provide the generic high-level purpose.

For each of the stages of development in the team's scope, the team should articulate the specific requirements that will be needed for approval, the specific tests that should be carried out, and the required state of the design at the time of approval. Articulating these items clarifies the specific design outcomes that should appear in the activity map for this specific project. This part of the customization process caters to the detailed and specific needs of the particular product being designed.

For teams working on products they're familiar with — such as the team that annually develops a new mobile phone — it will be relatively easy to articulate specific end-of-stage outcomes. This is because of their familiarity with the unique design patterns that exist for mobile phone development, and because they're likely to have a general idea for what the product will become, well before it is developed.

On the other hand, for teams working on products they're unfamiliar with, or on products that have never existed before, it may not be possible to be highly specific at the onset of the product development process. This is because there is still so much that is unknown about what the product will become that it's simply not possible to be specific. In such cases, it's acceptable for the team to start with generic end-of-stage outcomes, then update them when specific information becomes available.

When establishing the needed approvals, it is useful to evaluate the risks of failing to successfully meet the end-of-stage outcomes (such as failing to select a product architecture that can be

Figure 8.1: Scoped Development for Human-Powered Water Well Drill.

manufactured using the production resources of the client or its suppliers). For each of the risks, the team should establish an approval that will lessen or eliminate the risk (such as *early stage manufacturing process approval*).

Step 3: Identify Specific Intermediate Design Outcomes

The high-level information emerging from Steps 1 and 2 are useful for uniting the team, but it provides only minimal direction on how to proceed through the project. Step 3 solves this by establishing intermediate design outcomes (or subgoals) that if met would result in accomplishing the specific end-of-stage outcomes. To do this, the team breaks down each high-level outcome into smaller more manageable lower-level outcomes. *Decomposition* is a design tool that can help the team through this step; it is described in the Development Reference, under Decomposition (11.14). In the context of project goals, the team will use decomposition to ask *what lower-level things need to be accomplished in order to meet the higher level goals?*

Step 4: Choose Design Activities That Lead to the Outcomes

For each of the design outcomes resulting from Step 3, the development team chooses one or more design activities that alone or together are likely to result in the outcome. Many different activities or sets of activities can potentially result in a desirable outcome, so the selected activities are not unique. These activities, together with the design outcomes resulting from Step 2, can be arranged into an activity map that captures their dependency relationships.

For the most part, this portion of the customization process caters to the specific skills and preferences of the development

team. Part II of this book (the Development Reference) provides a collection of potentially useful design activities.

Step 5: Allocate Resources to Each Design Activity

For each of the selected design activities, the team allocates human and financial resources to execute the activity. One or more members of the product development team or its contractors are assigned the responsibility of carrying out the activity and accomplishing the outcome. Also, a portion of the overall project budget is thoughtfully allocated to the activity.

In the context of an activity map, these resources can be specified in a variety of ways. For simplicity, we have placed a name, and dollar amount next to each arrow in the activity map. For example, Figure 8.2 shows an activity map for one high-level activity for a consumer electronics project.

Step 6: Add Time Estimates and Extract the Project Schedule

A realistic estimate of the time it will take to complete each activity is made according to the human and financial resources allocated to it. Any unit of time can be allocated, but it is important that the same unit of time is used consistently. Generally, days and weeks work well for small-scale projects, and months or quarters work well for large-scale projects. Once estimates are made for each activity, the critical path can be found and the project schedule can be established. A method for determining the critical path is found in the Development Reference, under Critical Path Analysis (11.13).

* * * * *

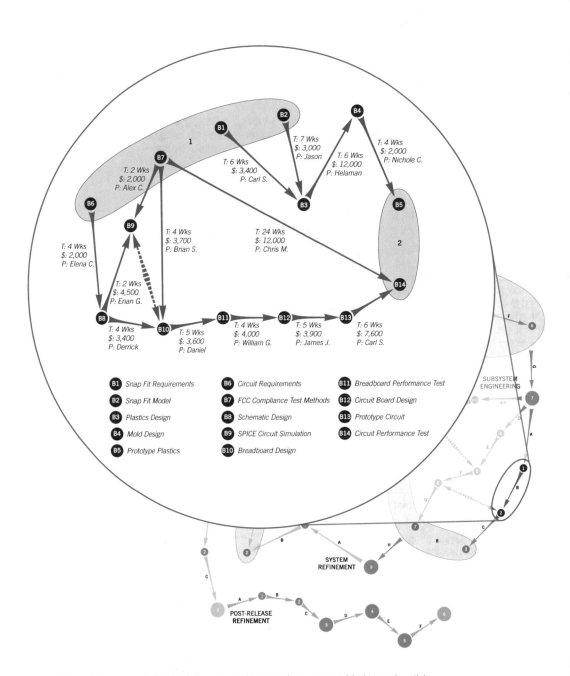

Figure 8.2: Activity map with financial, human, and temporal resources added to each activity.

For the most part, Steps 4–6 can be performed in any order. This means that after the specific intermediate design outcomes are identified (Step 3), the team could begin by allocating a portion of the overall resources and schedule to accomplishing the outcome. Within this limitation of allocated resources and schedule, the team can then choose activities and human resources that would be suitable.

Refinements and Adjustments to the Plan

If needed, the team can continue to decompose each subgoal into lower-level subgoals and repeat the Steps 2 through 6 where each subgoal is the subject of decomposition. It is important to decide how detailed to make the project plan. The plan should be (i) detailed enough to guide the product to the desired state of evolution with minimal waste, yet (ii) not so detailed that it removes the assigned individual's freedom to operate in the most efficient and effective way. The next section offers some suggestions that may help in deciding how much detail is needed.

Creating a customized project plan is an iterative process – meaning it is very unlikely that the first plan created will match the needs and constraints of the project. Planning can be effectively performed in a design iteration cycle (Section 2.2). The created plan needs to be evaluated and adjusted until it appropriately matches the needs and constraints. To adjust the plan, any one or more of the steps above are reconsidered and the impact of an adjustment carried throughout subsequent steps. The next section, on plan evaluation, offers some suggestions that help decide what adjustments might be needed. Generally, the team should be confident in higher level plans before creating more detailed plans.

Once established and in use, the project plan should be updated as necessary. We note that this will likely be frequently. By definition, because the plan is a detailed activity map, a change in one part of the map can affect another part. To avoid unanticipated problems, consider how subsequent dependent and interdependent activities are affected by any proposed map change before making the update.

8.2 Evaluating the Quality of the Project Plan

To evaluate the quality of a custom development plan, we recommend that each of the following questions be deliberately asked and thoughtfully answered about the project plan.

Does It Have an Appropriate Level of Detail?

This is a difficult — yet critical — question to answer. For more experienced product development teams, it is likely that fairly high-level plans are sufficient for the team to proceed. For less experienced teams, more detail will be needed simply because team members lack the experience to intuitively know what set of activities effectively lead to accomplishing the desired outcomes. The plan should contain just enough detail to result in efficient team work relatively free from wasted work due to misunderstanding the scope of the desired outcomes, the available resources, or the time constraints.

Does It Have Realistic Times Allocated?

It can be very difficult to estimate the time it will take to carry out an activity — especially interdependent activities. Generally activities take two to three times longer than most people estimate them to take. This can be very difficult for designers to deal with as there is always pressure to complete activities quicker than is estimated. When deadlines are fixed, the resources (human or financial) allocated to an activity, the scope of the activity, or both need to be adjusted (see Project Objective Statement (11.49) of the Development Reference).

Does It Have Realistic Costs Allocated?

Like time estimates, cost estimates are also difficult to establish. For many product development activities, these costs have to do with things such as material costs for prototypes, computing time with super computers, test lab fees, and payments made to interviewees or focus groups. In each of these cases, estimates can be made using experience, or speaking about the estimates with suppliers involved. These costs generally don't include the cost of human resources, but there would be nothing wrong with including it if desired.

Does It Have Appropriate Approvals to Minimize Risk?

Thinking about what might not go well during the process of evolving the product from its initial state of evolution to the final desired state is a valuable activity. The custom plan will be most successful when it is focused on creating approved designs as opposed to just any design. To that end, establishing the appropriate approvals is essential. For any industry there is generally a set of required approvals or standards that must be met, such as the UL 94 flame retardancy standard that exists for most consumer electronics. Be sure to understand the required approvals and standards that must be met for your specific project. These are sometimes unspoken because they are so commonly understood and therefore assumed.

Does It Capitalize on the Strengths of the People Assigned?

Remember that there are often various ways to accomplish the intermediate outcomes of a project. The plan should reflect activities the team is good at, or enjoys doing. If the plan requires an outcome the team is not capable of achieving alone, the plan should be coordinated with entities outside of the team such as a supplier. Also, within the team, the activities should be distributed such that the burden is appropriately shared and that individuals and teams are asked to carry out activities that align with their skills and interests.

Is It Useful?

The most important part of evaluating any plan is to ask if the plan is useful in helping the team know what needs to be done, by what time, and with what resources. One indication of this is if the team uses and updates the plan regularly. Generally the creation of the plan will be largely driven by one individual on the team, such as the team leader. It's our belief that if the one creating the plan is thoughtful about its creation and diligent in updating and emphasizing of the plan, the plan will indeed be of great use to the team.

8.3 Using the Customized Project Plan

The custom project plan represents what the product development team will do to advance the design through the stages of development, but it does not represent *how* they will do it. For example, imagine two product development teams with identical project plans for evolving two products for the same market. How the team progresses through the map — with what level of care and with how much enthusiasm — has a large impact on the cost and quality of the product development effort.

In order to ensure that the resulting product is both desirable and transferable, the project should be both well planned and well managed.

In this section, we describe project planning and management as the two halves to the common maxim "Plan your work, then work your plan." The first half is accomplished through project planning; the second, through project management.

Without project planning, the approach for transitioning the design through the stages of development is not clearly defined. This makes it difficult, if not impossible, for all team members to pull together to accomplish the project goals. Without project management, there is no guarantee that the team activities will be coordinated to achieve the plan. Instead, the project may end up at an entirely different place or take significantly longer than planned for.

We now discuss both Project Planning and Project Management in more detail.

Project Planning

The goal of project planning is to completely define the whole activity map in the context of the product and project needs, development resources, and schedule implications of the project. This typically includes defining the following:

- Scope
  - Starting state
  - Ending state
  - Breadth of development project
- Schedule
  - Calendar time available to complete the project
  - Milestones between the start and end
- Resources
  - Available money
  - Available equipment, tools, software, etc.
  - Available people
  - Stakeholders
  - Team members
  - Approvers
- Trade-off criteria
  - Some understanding of what will most likely be adjusted in the event of unforeseen difficulties. Do we adjust the ending state of the project (scope), the amount of resources allocated (resources), or the amount of calendar time allocated (schedule)?

The top-level project plan should be approved by the client. Once it is approved, the team can move ahead more confidently, knowing the client has some ownership in the plan. Any changes to the plan should be openly discussed and validated by the client. The project plan will generally include a project contract and schedule.

Project Management

The goal of project management is to ensure that the project plan is followed, or that any necessary deviations from the plan are carefully and thoughtfully executed. A secondary goal of project management is to be able to give an accurate status update at any time. Status should include project state, resources used and remaining, and time state (ahead of schedule, on schedule, or behind schedule).

In general, there is no formal test or validation of project management. On a regular basis however, the current performance of the team should be compared with the planned performance.

In many cases, there is no formal deliverable associated with project management. But an accounting of the project status should be tracked by the team for self-evaluation, and reported to the client on a regular basis.

Useful Tools for Project Planning and Management

Gantt charts, activity maps, critical path identification, PERT charts, work breakdown structure, project planning software.

Things to Watch Out for

Avoiding planning
At the beginning of the project, it feels like not enough is known to produce a good plan. It can be tempting to wait until more is known. This can lead to a state of perpetual working, but with no plan. It is far better to begin with a tentative plan and then refine it as more information is obtained.

Perpetual planning
It is possible to get so involved in the planning process that the team never moves to a whole-hearted execution of it. It is vital to move to execution at the appropriate time. Deciding the appropriate time is a matter of judgment.

Ignoring the plan
For some teams, the plan can unfortunately be viewed as a hurdle to get over, or a hoop to jump through. In this case, once the plan is created, the team (sometimes unconsciously) moves ahead as if they had never made the plan!

Making the planning documents the goal
The objective of the project is to change the state of the design. The plan is a means to this end. In error, the team can make project management the goal, instead of the means to the end.

Pretend plans and statuses
Unfortunately, some will feel the pressure to tell the stakeholders what they want to hear, instead of sharing reality with them. To be effective, the plan and the status of its execution must reflect the reality of the project.

We believe that one of the greatest variables at our disposal in developing a product is the product development process itself. To that end, we have presented a 6-step method for creating a customized product development process and have shown how it can be more fully detailed into a unique project plan.

8.4   Human-Powered Drill Example

Let us now examine how the customization process unfolded — at the onset of the product development process — for the human-powered drill. We note that as development continued for this project, the plan presented here became more and more detailed as more and more became known about the selected architecture and the unique subsystem engineering challenges.

Step 1 (Example): Establish the Project Scope for the Human-Powered Drill

Through thoughtful discussion with the client, the team gathered essential scope information, which they crafted into the following project objective statement:

> Design, build, and test a human-powered drill that reaches underground potable water at depths of 250 ft in all soil types by March 25, 2011 with a prototyping budget of $2,800 USD and for less than 1,700 man hours of development.

Because the team needed to produce a complete system for field testing, and recognizing that some work had already been done, the team bracketed the development to be from the middle of opportunity development to the end of the system refinement stage of product development. The scope of development for the drill is illustrated in Figure 8.1 and a set of high-level development milestones is provided in Table 8.1.

Step 2 (Example): Articulate Specific End-of-Stage Outcomes for the Human-Powered Drill

From Step 1, the team knew that it would transition the design through the first four stages of development. Using Tables 4.1, 5.1, 6.1, 7.1, the team articulated — as specifically as it could be given the novelty of the product — the specific end-of-stage outcomes for the human-powered drill. Table 8.2 shows the team's result for Step 2.

Step 3 (Example): Identify Specific Intermediate Design Outcomes for the Human-Powered Drill

As the project started, the team identified a few intermediate design outcomes that would help it achieve the end-of-stage outcomes. Later, as more become known about the drill, additional specific intermediate design outcomes were identified and added to the plan.

For the opportunity development stage, to be able to acquire a deep understanding of the product expectations, the team identified a few specific outcomes to pursue: (i) understanding WHOlives.org's wishes for the project, (ii) a summary of information gained from well driller

Development Milestone	Target Date
Opportunity development stage complete	30 Sep 2010
Concept development stage complete	30 Nov 2010
Subsystem engineering stage complete	31 Jan 2011
System refinement stage complete	25 Mar 2011
Final, pre Tanzania, design review	15 Apr 2011
Field testing in Tanzania	15 May 2011

Table 8.1: Development milestones for the human-powered water well drill project.

interviews, (iii) a summary of information gained from internet research, and (iv) a summary of information gained from benchmarking other competitive drills.

For the concept development stage, the team decided that they needed an understanding of operator fatigue (Outcome 2 in the concept development stage of Figure 8.3), labor ergonomics (Outcome 3), rate of cuttings removal (4), chip settling in a slurry pond (5), pump pressure (6), mechanical advantage (7), and manufacturing cost (8). Low-cost evaluations of many prototypes were chosen to achieve Outcomes 2 through 5. Low-cost analytical models were chosen to achieve Outcomes 6 and 7, and simple cost-estimating models were chosen to achieve Outcome 8.

For the subsystem engineering stage, the team chose to have each subsystem tested thoroughly and under harsh conditions — even though the team did not know what those subsystems would be when they identified this intermediate outcome. The team hoped that it could use the results from each test as a way to identify problems and push to correct them before the next drilling test. The subsystem tests are represented by compound Outcomes 14 and 15 in subsystem engineering in Figure 8.3.

For system refinement, the team planned to have the entire drill system tested in Utah. They also planned to have the entire drill system tested in Tanzania. These tests are compound Outcomes 5 and 10 in system refinement in Figure 8.3.

Step 4 (Example): Choose Design Activities That Lead to the Outcomes for the Human-Powered Drill

For each outcome, one or more design activities are chosen to lead to the outcome. For the outcomes listed in Step 3, the team decides to do the following.

For the opportunity development stage: (i) interview the client (WHOlives.org), (ii) craft and revise the project objective statement, (iii) interview well drillers, (iv) do internet research, (v) benchmark competitive products, and (vi) discuss/brainstorm/establish product expectations based on interviews, internet, and benchmarking data.

For the concept development stage: (i) create and test a low-fidelity prototype to predict user fatigue, (ii) create and test a low-fidelity ergonomics prototype, (iii) create a simple test to measure rate of dirt removal and run the test, (iv) create and test a chip settling prototype, (v) create a predictive measure of mechanical advantage, (vi) develop a predictive model of required pump horse power, and (vii) create a preliminary cost model. These activities are shown a B–H, respectively, in Figure 8.3.

For the subsystem engineering stage: (i) test each subsystem under harsh conditions to expose weakness, (ii) refine each subsystem to strengthen weaknesses identified during tests, and (iii) retest each subsystem to prove it has been strengthened.

For the system refinement stage: (i) test the entire integrated system in Utah, and (ii)

Table 8.2: Specific end-of-stage outcomes for human-powered drill.

Opportunity Development:
Requirements (required content at the end of stage):
Market requirements, performance measures, requirement measure relationships, and ideal values (sections A-D) of the human-powered drill requirements matrix
Tests (required content):
None
Design (required content):
None
Validation for approval:
Do the ideal values seem reasonable to WHOlives.org (for better or for worse, WHOlives.org was chosen as the market representative for the human-powered drill project)? Does WHOlives.org know of any missing performance measures or market requirements that are not captured in the requirements matrix? Does WHOlives.org believe the importance values are correct?

Concept Development:
Requirements (required content at the end of stage):
Target values for the drill performance in the system-level requirements matrix. Sections A-D and target values for subsystem requirements matrices.
Tests (required content):
Modeling methods used to predict performance of the drill concept and test methods for prototypes used to measure performance of the drill concept.
Design (required content):
Overall concept for the drill. Decomposition of drill into subsystems. Interfaces between subsystems.
Validation for approval:
Predicted and measured performance of the drill concept is consistent with the established target values. WHOlives.org is happy with the concept as expressed in the prototypes and sketches.

Subsystem Engineering:
Requirements (required content at the end of stage):
Predicted and measured performance values in the drill's subsystems requirements matrices. Predicted performance values of the drill in the system-level requirements matrix.
Tests (required content):
Updated methods used to predict and measure the drill and subsystem performance.
Design (required content):
Design for each of the drill's subsystems. This will include a bill of materials, engineering drawings of custom parts, purchasing specifications for off the shelf parts, system assembly instructions, piping diagram for cutting fluid.
Validation for approval:
Does the measured performance meet or exceed the target performance? Is it within the acceptable limits? Where possible WHOlives.org will critique prototypes and performance data resulting from this stage.

System Refinement:
Requirements (required content at the end of stage):
Measured drill performance values in the requirements matrix. Refinements to predicted and measured performance values for the drill's subsystems in the subsystem matrices.
Tests (required content):
Updated methods used to predict and measure the drill and subsystem performance.
Design (required content):
Refined design for the entire drill.
Validation for approval:
Does the measured performance meet or exceed the target performance? Does WHOlives.org and other market representatives in Tanzania find the product desirable?

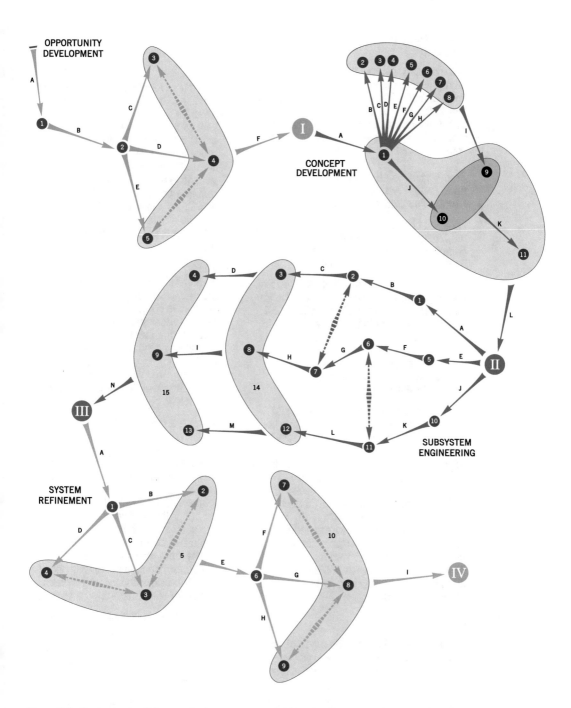

Figure 8.3: Customized activity map for human-powered drill, reflecting specific intermediate design outcomes for the drill.

test the entire integrated system in Tanzania.

Once chosen, each of these activities can be represented in the design activity map. For readability (based on the constraints of a layout), we show this for only the opportunity development stage. The activity map is shown in Figure 8.4, along with the resource allocation from Step 5.

Step 5 (Example): Allocate Resources to Each Design Activity for the Human-Powered Drill

For each chosen activity in the activity map, resources are allocated to its completion. This includes human, financial, and temporal resources.

Figure 8.4 shows the result of this step for the opportunity development stage. Notice that stage completion date from Table 8.1 has been included in the figure. This is to facilitate Step 6.

Step 6 (Example): Add Time Estimates and Extract the Project Schedule for the Human-Powered Drill

A project schedule for the entire project can be extracted once resources have been allocated to all the activities in the activity map. Critical path analysis (see the Development Reference) can be used to do this. For the activity map shown in Figure 8.4, the critical path analysis indicates that all activities except *interview well drillers* are on the critical path. This means that (based on the allocated time resources) no activity, except *interview well drillers*, can be delayed without consequently delaying the 30 September 2010 milestone. The extracted schedule is shown in Table 8.3.

Discussion

Various observations can be made about this example that may help you as you create a custom project plan for the first time.

- The resulting plan is unique to the project. Figure 8.3 reflects specific outcomes and activities that apply only to the drill. Also Figure 8.3 conveys

the fact that the human-powered drill project only covered the first four stages of development.

- The plan evolves as more information is gained. In this example, we chose to describe the custom plan as it existed very early on for the team — at the onset of the product development process. As a result, the plan is somewhat specific but still very generic. The plan will evolve as the requirements, tests, and the design evolve. As a result, it's OK that the plan be a detailed plan in the near future, and a less detailed plan in the distant future.

- There are various acceptable plans. Numerous plans could have been created to achieve the outcomes. Any alternative choice for the intermediate design outcomes, the activity, the estimated duration, the finances allocated, or the person assigned to lead the activity would constitute a different plan. Various acceptable plans can result from the flexibility available.

- There are many unacceptable plans. While there are various acceptable plans, we must also recognize that there are many unacceptable plans that will not lead to meeting the design outcomes. For example, choosing to interview the client for the first time only after the performance expectations have been discussed/brainstormed/established by the team would be a mistake since the performance expectations need to reflect the expectations of the client and other stakeholders.

- The best plan will have the following characteristics. It will:
  - Produce the desired outcomes while simultaneously meeting the desired budget and milestone goals.
  - Make the best use of product development resources (human, financial, and temporal).

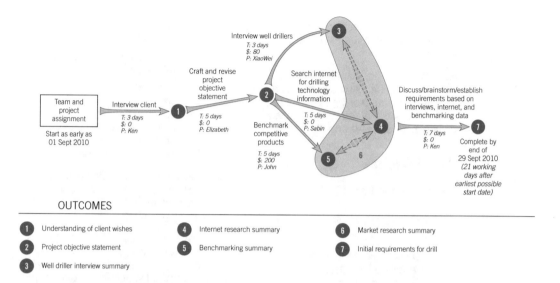

OUTCOMES

1 Understanding of client wishes   **4** Internet research summary   **6** Market research summary

2 Project objective statement   **5** Benchmarking summary   **7** Initial requirements for drill

3 Well driller interview summary

Figure 8.4: Resource allocation to the human-powered drill project (opportunity development stage only).

Activity	Scheduled Start Date	Required Completion Date
A. Interview WHOlives.org	01 Sep 2010	03 Sep 2010
B. Craft and revise project objective statement	07 Sep 2010	13 Sep 2010
C. Interview well drillers	14 Sep 2010	20 Sep 2010
D. Do internet research	14 Sep 2010	20 Sep 2010
E. Benchmark competitive products	14 Sep 2010	20 Sep 2010
F. Discuss/brainstorm/establish product expectations	21 Sep 2010	29 Sep 2010

Table 8.3: Development schedule for the human-powered drill (opportunity development stage only).

- Facilitate a common understanding (among team members) of specific design activities, their outcomes and relationships.

- Be detailed enough to guide the product development process, yet flexible enough to allow individuals to apply their individual strengths to the project.

8.5   Summary

Every project needs its own custom product development plan. The team's opportunity to construct that plan in such a way that maximizes efficiency and effectiveness

means that the product development process itself is one of the greatest variables at their disposal.

Six steps can be followed to develop a custom plan for any specific project. Those steps are (1) establish the project scope, (2) articulate specific end-of-stage outcomes, (3) identify specific intermediate design outcomes, (4) choose design activities that lead to the outcomes, (5) allocate resources to the each design activity, and (6) add time estimates and extract the project schedule.

Project plans need to be made and managed. Project plans must have an appropriate level of detail, have realistic timeframes, have realistic costs, have appropriate approvals to minimize risk, capitalize on the strengths of the team, and be useful for managing the work on the project.

Key risks of project planning include avoiding planning, perpetual planning, ignoring the plan, making planning (rather than evolving the product) the goal, and making pretend plans and statuses.

8.6   Exercises

Test Your Knowledge

T8-1 List the steps in the process for customizing the project plan.

T8-2 List six questions that can be asked to evaluate the quality of a customized project plan.

T8-3 List five potential pitfalls in project planning and management.

T8-4 List the three major elements of a project plan that may need to be adjusted when unforeseen difficulties arise.

Apply Your Understanding

A8-1 Give an example of a project where you underestimated the time required for completion. What is the approximate ratio of the actual time to the estimated time.

    a) What criteria would you expect to use when assigning human resources to particular design activities?

    b) What criteria would you expect to use when assigning financial resources to particular design activities?

    c) In general, how do you expect to know when a design activity is complete?

A8-2 Consider a development project you are working on now.

    a) What stages of development does your project cover?

    b) What stage of development is it currently in?

    c) What is the next development milestone?

    d) How much time and money do you expect it will take to advance the design to the next milestone? Do you have enough time and money available to reach the milestone?

A8-3 Describe the difference between a customized product development process and a customized product development plan.

A8-4 The order of steps 4-6 of the process for customizing a product development plan is interchangeable. How would your thinking differ if you completed these steps in the order 6-5-4 instead of 4-5-6?

A8-5 Using the critical path method, find the critical path for the activity map shown in Figure 8.2.

CHAPTER 9

Seven Ways to Become a Better Designer

This book has been mostly about the mechanics of product development; the concept of product evolution, its stages of development, and a process for planning design activities that promote that evolution. All of these things are meant to help you become better designers. At some point, however, the mechanics of product development fade into the background and our natural design instincts and experience take over. That's not to say the mechanics are ignored, it simply means that our design intuition becomes well-aligned with the fundamental mechanics of the process.

The goal of this chapter is to share some non-mechanical things that will help you become better designers.

Over the years we have observed many new designers become great designers as they have done the things described in this chapter. We believe that these seven things (Sections 9.1–9.7) will help all of us become better designers and find greater fulfillment and success in our product development.

9.1 Expect Challenges

Simply stated, product development is difficult — very difficult. Great designers understand this and work to minimize difficulties — but they are not overly surprised or frustrated by the challenges of product development.

What makes product development so difficult?

One of the greatest challenges of product development is *uncertainty*. What is known about the design, or even the opportunity, is uncertain and incomplete especially at the beginning of the product development process. Great designers seem to embrace this uncertainty and learn to work with the information that is available, filling in with appropriate assumptions and updating their thinking and approaches as new information is discovered.

If you're expecting the problem to be neatly prescribed, or to develop a solution without bumps along the way, you'll be disappointed and frustrated. Embrace the

© Springer Nature Switzerland AG 2020
C. A. Mattson, C. D. Sorensen, *Product Development*,
https://doi.org/10.1007/978-3-030-14899-7_9

uncertainty and ambiguity and consider it your job to thrive in that environment.

Another challenge is that there are almost always competing design objectives. It's just not possible to give the client the least expensive and the most reliable vehicle on the road at the same time. Therefore the designer has to strike a balance between competing objectives — keeping in mind the interests of everyone involved. This means choices have to be made — tough choices that require you to give up some good performance in one area to gain better performance in another. Great designers know this will happen and are ready to tackle those competing objectives without being surprised or frustrated by them.

Great designers seem to deal really well with the challenges of time. They are acutely aware of the time remaining to complete a project. They don't ignore this universal constraint. At the same time, they realize that there is never enough time and that the design can always be improved further with more time. But they never use this as an excuse to do a poor job; they develop the best possible product in the time available. They learn to negotiate with the client to refine the scope of the project as time goes on.

Working on a team is a challenging part of product development. Almost all teams go through a phase of conflict where they are learning to work with one another. Great designers are flexible when working with others, yet firm in their dedication to developing the best product. To minimize conflict, the greatest designers seem to fight for the needed design outcomes while conceding that a number of different activities could lead to that outcome. Knowing that there are many *right* ways to do something reduces many team conflicts. When there are conflicts, great designers try to minimize the time lost by seeking to resolve them quickly and effectively.

Product development can also be difficult when we lack a particular skill that is needed to carry out the project. Great designers are aware of their own skills and know when and where to quickly learn something new or how to find help from someone with greater expertise.

9.2   Be Mindful of the Design

Great designers understand that the final design will emerge little-by-little in an evolutionary way. Ideally, everything the designer does advances the design to a higher state of desirability or to a more ready state of transferability. The greatest designers are always aware of (i) the current state of the design (in its evolution), (ii) the final desired state of the design, *and* (iii) how their current actions are leading to the final state. Great designers use this information to take control of the design and facilitate its evolution. This is so different than the designer who passively lets the design become what it becomes with only minimal coaxing.

As an example of minimal coaxing, we've seen some designers build analytical models solely for the sake of analysis, without a deep understanding of how those models will impact the design. But great designers don't do this. They are quite aware of what they're working on and precisely how the design might advance if their work is successful.

9.3   Be on the Team

In any project, there are team objectives and personal motives — and for some people these two motives can be at odds. But not for great designers. They align their personal motives with the team objectives so that their efforts support not only their personal interests but also the objectives of the team. This is not to say they become submissive and robotic. Great designers have and share their opinions, but they are not opinionated.

When good designers consider the full set of product development stages, they see that it is very unlikely that a single person will possess all the skills alone to evolve the design through all of the stages. Great designers are aware of their own limitations and recognize the strengths that others bring to the team. They truly believe that the best solution is the team solution — not the solution from any one individual.

DIVERGE   **CONVERGE**

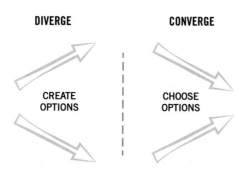

CREATE   **CHOOSE**
OPTIONS   **OPTIONS**

Figure 9.1: The pattern of diverging by creating alternatives from which to choose and converging by making choices from among these alternatives is one of the hallmarks of a great designer. Figure after Brown (2009).

Great designers also know how to give and receive feedback in a team setting. The feedback they give seems to be based only on getting a better design, not on criticizing fellow team members. They are willing to see other people's views and are slow to be defensive to critiques.

9.4   Diverge Then Converge

Great designers know that before converging on a design there must exist a period of divergent exploration (see Figure 9.1). During this period, many diverse (and potentially wild) options are briefly considered. By considering these options, the designer forms a sort of design space in the mind that allows him or her to envision where the best spots of the design space are. Those best spots are then further explored.

Great designers don't believe that the best design is the first design that pops into one's mind; and in the rare case where it is, it is only considered the *best* once many other diverse designs have been considered.

At the same time, great designers understand when the time for divergence has past and the time for convergence takes over. Even though divergent thinking is fun and exhilarating, and there are so many uncertainties that make it difficult to converge on a single design, great designers are not afraid to tackle this head on. They understand the importance of making decisions that eliminate options and have the discipline to do it.

9.5   Embrace Multifidelity Models and Prototypes

Many novice designers believe that predictive models and physical prototypes need to be quasi-perfect to be valuable. The best designers don't believe this at all. You can take a shoebox and drop it in front of a great designer and call it a defibrillator and the designer will understand and appreciate that it is a low-fidelity prototype. She'll immediately begin thinking about if this is the right size for a defibrillator, how heavy it might be, how it would be used, and what additional design information would need to be defined before a higher-fidelity prototype could be built. Great designers value these low-fidelity prototypes because the amount of learning and thought provoking imagination is very high compared to the cost of making it.

At the same time, great designers value high fidelity prototypes, knowing that the cardboard box is inappropriately crude for seeking feedback at a trade show. For that, great designers prepare a different prototype. Importantly, great designers do not value only low-fidelity prototypes or only high fidelity prototypes, they value both depending on the goal of the prototype.

The same is true with predictive models. A low-fidelity prediction that points the designer to the main variables that can be changed to get a desirable outcome is very valuable. The best designers know when best to use both low and high fidelity models.

9.6   Validate Assumptions

Many assumptions will be made during the product development process — mostly as a way of avoiding stagnation. Great designers know, however, that every assumption left unvalidated is a seed that can potentially lead to failure.

Assumptions come in every flavor during product development. It may be an assumption about friction or inertia. Or it might be an assumption about the cost of components purchased in bulk. Or the assumptions about manufacturing labor rates. But the assumptions most likely to lead to failure are assumptions about what the client and market want. Novice designers often use their own preferences and paradigms as a substitute for those of the client and market. This is a mistake great designers don't make. Almost always this mistake leads to products that are loved by the designer, but not so loved by the client or market.

One of the realities of product development is that the designer is a sort of facilitator, facilitating the process of evolution. He or she is the connection between the client and the product that he or she wants. Great designers listen to the client and the market. They ask thoughtful questions without leading the client to a hoped-for answer. They follow up with more questions — seeking a deeper understanding — knowing that the questions are also helping the client discover what he or she really wants.

Great designers consider data to be a valuable part of validation and are doubtful when key decisions are made without sufficient data. They appreciate data that comes from both quantitative and qualitative tests because he or she knows that both kinds of tests may be needed to successfully develop the product.

9.7   Always Have a Back-Up Solution

Many of the things we design are not going to work at first the way we thought they would. There are just too many unknowns until we try it. In anticipation of this, great designers always have a back-up solution brewing in the back of their mind. When things don't work out as planned those back-up solutions have already been thought about to some degree and can quickly come to the rescue of a previous failed attempt. Most often, these back-up solutions are small changes to the design that improve the chances of desirable functionality, not radical changes that are also unlikely to work on the first try.

9.8   Beyond These Seven Suggestions

Becoming a better designer is a lifelong pursuit. This doesn't mean you can't become a great designer now — in fact you can. But in this broad field of design — one that requires skills in creativity, analysis, craftsmanship, ethnography, manufacturing, statistics, and on and on and on — there is room for each of us to become better.

While we have listed seven ways we have seen many new designers become better, you may already be great at these seven things. If that is the case now, or when that becomes the case, we encourage you to find a few new areas to develop competence in while maintaining your deeper competency in the area that people have come to rely on you for. We believe that doing this will not only bring you greater fulfillment in your product development pursuits, but it will put you in a better position to serve the client and help him or her deliver a product that the market truly loves.

Apply Your Understanding

A9-1 Write a short paragraph articulating your competencies others have come to rely on.

A9-2 Develop a specific plan to increase your competence in one of the seven ways to become a better designer.

A9-3 List a few back-up solutions you have for a particular project you are currently working on.

A9-4 List some unanticipated problems that have appeared in a product development project you are familiar with, and how those problems were overcome.

A9-5 Describe a time when you abandoned a product development activity because you realized it was not contributing to the evolution of the design.

A9-6 Tell how a clear focus on evolution caused a product development effort to move forward rapidly and efficiently.

A9-7 What suggestions would you give to a novice designer about receiving feedback on product development work?

A9-8 What actions can be taken to align personal motives with team objectives?

A9-9 How does the diverge-converge process in this chapter relate to the Controlled Convergence method described in Controlled Convergence (11.10) of the Development Reference?

CHAPTER 10

Conclusion

Looking back now at my[1] first trip to Taiwan — the one where I worked on the docking station — I can see that it changed me. I can even pin-point where it happened and how I changed. I was on the flight, in a comfortable seat on the upper deck of the Boeing 747. As I sat there reading a product development textbook trying to get ready for the wave of product development activities that would happen when we touched down, I remember thinking *I can do this!* I gained confidence that product development was not only for the elite, but that I could do it too and do it well. Even though I was in my last year as a university student, I transitioned — at that moment — from being student of product development to being a product developer.

The goal of this book has been to help you transition into an effective product developer, whether you're designing docking stations or landing gear for the 747. To that end this book has introduced exactly what I needed to know — when I was on that plane — about how products are created. To summarize, these are the key points we've raised in this book:

- Products are manufactured by a production system, not the development team. This means that

the development team's job is to create a design, rather than to create a product.

- Successful designs are desirable and transferable. Transferable means that the production system can use the design to manufacture the product as the development team intended with no outside communication. Desirable means that the product manufactured in accordance with the design will be desired by the market.

- The design evolves throughout the product development process. From its initial state, the design evolves until it has a fully detailed, transferable design that can be released to a production system. After being introduced to the market, the product is periodically refined throughout its lifetime.

- Effective product development evolves the design in stages. Each stage has a unique focus in the product evolution. Later stages are not begun until the results of earlier stages have been approved as meeting the requirements for desirability and transferability.

- Three things evolve during the product development process: requirements, tests, and the design. All three are

---

[1]This is Chris's personal experience.

© Springer Nature Switzerland AG 2020
C. A. Mattson, C. D. Sorensen, *Product Development*,
https://doi.org/10.1007/978-3-030-14899-7_10

necessary to achieve a desirable and transferable design.

- Models and prototypes are an essential part of developing engineered products. Rather than evolving, models and prototypes are created based on the design at the time the model or prototype is created. Models and prototypes are used to predict and test performance, and to obtain market validation of the desirability of the design.

- Design evolution happens in small increments, where proposed changes are implemented and evaluated. If the changes are desirable, they are adopted as part of the transferable design.

- There are crucial design skills for effective product development. These skills include planning, discovering, creating, representing, modeling, prototyping, experimenting, evaluating, deciding, and conveying.

- Product development is carried out by teams, and coordination among team members is essential for efficient product development. An activity map provides a visual representation of the coordination plan. Effective project management coordinates the activities of individual members to follow the coordination plan.

- The best product development process must be customized for an individual development project and team. Only by customizing the process to the specific development situation can the efficiency and effectiveness be maximized.

- At its core, design is a human activity. The best designs are likely to be created by the best designers. Seven suggestions for becoming a better designer are (1) expect a challenge, (2) be mindful of the design, (3) be on the team, (4) diverge then converge, (5) embrace multifidelity models and prototypes, (6) validate assumptions, and (7) always have a back-up solution.

We believe that the points articulated above will help anyone, novice or expert, improve the effectiveness of their product development work. But mechanically following these points and other principles taught in this book will not produce the desired outcomes. Product development is a creative endeavor, and requires emotional and intellectual investment from the development team.

As you are learning, product development is fun, challenging, difficult, exciting, stressful, and exhilarating. Developing successful products requires the best efforts you have to give. We encourage you to rise to the challenge, and find the joy of creating products that meet human needs and make the world a better place.

Happy product development!

Part II

Product Development Reference

CHAPTER

11

Product Development Tools and Techniques

The product development reference in this chapter provides a brief introduction to various tools and techniques that are useful in product development.

Each entry in the development reference has a title (name of tool or technique) and short statement about why that tool is useful, followed by basic information and how-to guides. Many entries provide sources for further reading.

To facilitate finding information in the development reference, the tools are ordered alphabetically. Further, the colored tabs on each entry indicate the stages of development for which the tool is often helpful. Related entries can be found by referring to the See Also section of each entry.

© Springer Nature Switzerland AG 2020
C. A. Mattson, C. D. Sorensen, *Product Development*,
https://doi.org/10.1007/978-3-030-14899-7_11

11.1 Basic Design Process

Develop a refined solution that meets real needs

The basic design process is a five-step process used to help designers be effective. Nearly all types of design can be generalized by the basic process shown in the figure. When carried out thoughtfully, the process leads to good designs.

How to do it

1. Understand the need: Start by trying to understand the problem that needs to be solved. Learn about who has the problem and what their needs and wishes are for the design. Engage in research and discussion to learn. Try observing users to learn about the problem.
2. Explore concepts: Once you understand the main problem, you can start exploring concepts to solve that problem. It's helpful to explore the possibilities by considering lots of diverse concepts. With many options on the table, you can confidently converge on the concept you believe has the most potential to solve the problem.
3. Define the design: Add details to the selected concept, giving it a more precise definition. Add enough detail so that you can develop a model of the design and predict its behavior. Experimenting with those models will help you define a good design and show others how well it will work. Experimenting with prototypes can also be useful as it will help you understand phenomena not captured by your predictive models.
4. Test the design: Test your design to explicitly see if it meets the identified needs. Try making and testing prototypes to better learn how well your design works in a real life setting. Evaluate the results of your testing and the quality of the design. Share your prototypes with potential customers to obtain their evaluation of the design.
5. Refine the design: If the tests reveal that improvements should be made, refine the design by returning to one or more of the previous steps. Expect the design to need refinement; it is unlikely for anyone to create the best design with just one pass. Eventually, after refinement, the tests will reveal that the design sufficiently meets the needs.

Common design approaches

In addition to the steps listed above, the basic design process includes common design approaches. These approaches can be applied to a variety of specific activities, which should be chosen to meet the needs of an individual project.

Engage: It is important for designers to see other perspectives, different from their own. To do this, designers will need to engage with the outside world. They will talk to people, listen to their story, and get to know their needs and wishes for a solution. They can also become familiar with similar needs and how they're solved in a different setting. This can help establish the constraints/limits of a new solution.

Observe: What people say is often different from what they do. Designers who work based solely on what their customers say often miss important customer needs that make the difference between a product customers truly love and one that merely satisfies them. To observe, pay attention to body language, non-verbal audible cues, and to responsiveness. Also pay attention to the physical surroundings, including wear patterns and work-arounds.

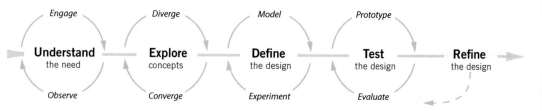

Graphical representation of
the Basic Design Process

Diverge: When designers diverge, they explore many different concepts. Designers do this to search for concepts that are better than the initial ones that come to mind. To diverge, think about concepts that are noticeably different than the ones already considered. Try combining two concepts together into a new concept. Or try creating the *inexpensive one*, or the *rugged one*, or the *eco-friendly one*. Don't worry too much about feasibility at this point; infeasible concepts can be useful stepping stones that can/should often lead to feasible ones.

Converge: When designers converge, they narrow their concept set down to just one that will be further developed. To converge, eliminate concepts that are infeasible. Identify the few concepts that are most likely to meet the needs identified earlier in the process. Intuition plays a large role in convergence. Develop your intuition, and your confidence in following it. The more complex the problem is, the more valuable the systematic convergence activities become; they can lead to non-intuitive desirable concept or in some complex cases may be the only way to find a feasible concept.

Model: When adding detail to the selected concept, designers will choose specific details such as a specific material or specific geometric dimensions. Predictive models are often invaluable when doing this. With relatively little cost, many variations of a concept can be considered and experimented with before defining which specific values to choose.

Experiment: With predictive models, it is easy to experiment with the design and improve it before it is built. Simple experiments can be really useful, especially when you have experience that improves your intuition. More complex experimentation methods using designed experiments and/or numerical optimization techniques can also be useful for defining some designs.

Prototype: Whenever possible, designs should be physically built and tested. The reality is that predictive models are only approximations of the physical world. Prototypes expose designs to real conditions, which often help designers recognize weakness in the design. Both rough and refined prototypes have a role in design. Designers generally choose the least expensive prototype that will sufficiently indicate how well the design meets the needs.

Evaluate: A thoughtful analysis of the design is needed to determine if the design is good enough or if it needs to be refined by repeating one or more of the previous parts of the basic design process. Quantitative and qualitative evaluations will most likely be needed to evaluate the performance of the design, its potential for improvement, and the cost (time and money) to improve it.

See also

Development Reference: Brainstorming (11.5); Catalog Search (11.7); Delphi Method (11.15); Design of Experiments (11.18); Experimentation (11.26); Focus Groups (11.31); Internet Research (11.33); Interviews (11.34); Method 635 (11.35); Observational Studies (11.40); Prototyping (11.50); SCAMPER (11.58); Scoring Matrix (11.59); Screening Matrix (11.60); Surveys (11.65);

Basic Design Process

11.2 Benchmarking

Understand and use the best available ideas

Benchmarking is a process of developing a thorough understanding of competitive products. These products are evaluated to determine how well the market likes the product, key strengths and weaknesses, the performance of the product on the performance measures, the concepts used in the product, and the estimated manufacturing cost of the product.

Market benchmarking

Market benchmarking is the process of obtaining ratings from the market or market representatives about how well competitive product meets the market requirements.

Market benchmarking involves obtaining information from the market or from market representatives. The highest quality information is directly obtained from the market. Information about purchase decisions is the most accurate information available, but it does not generally include reasons for the purchase decision that was made. Market representatives can supply reasons for purchase decisions.

Interviews, focus groups, and surveys of customers of competing products can provide important information. The techniques for eliciting customer information that are described by Ulrich and Eppinger (2012, Chapter 5) are useful in the benchmarking process.

Magazines and trade reports can provide market benchmarking information.

A relatively new source of market benchmarking information is online reviews. Many products have reviews that indicate likes and dislikes for products that have been purchased. Coupled with market share data, this can provide insight into the market approval of products.

Technical benchmarking

Technical benchmarking is the process of determining the level of performance of competitive products in each of the performance measures defined in the opportunity.

Technical benchmarking involves measuring the performance of competitive products relative to the performance measures. This provides an excellent opportunity to evaluate your testing procedures for the performance measures. Each key competitive product should be evaluated.

Using market and technical benchmarking

The combined results of market and technical benchmarking are used to determine ideal and marginal values for the performance measures. Because technical benchmarking has been completed, the team understands how each benchmarked product performs on each of the performance measures. When correlated with the market rating data obtained from market benchmarking, the value the market places on various levels of performance can be inferred.

When using benchmarking to help identify ideal and marginal values for performance measures, it is important to understand that the competitive products are generally improving over time. This means that if your development timeline is more than a few months, you should allow for the continued increase in competitive product performance, and you should aim for performance that is better than the current competitors.

Further Reading

The American Society for Quality describes the use of benchmarking in developing requirements matrices: http://asq.org/learn-about-quality/benchmarking/overview/tutorial-building-house-of-quality.html

Design benchmarking

Design benchmarking is the process of understanding how competitive products are designed to meet the market requirements and the performance measures.

Design benchmarking involves obtaining competitive products, disassembling them, and developing an understanding of what each feature in the product does. During design benchmarking, it is also common to estimate the cost of each of the components in the design.

Design benchmarking is often used to establish a minimum baseline for the design team. If the team cannot find a design that is better than the best competitive design, the team is required to use the competitive design. Because designers generally dislike using other designs, this serves as a strong motivation for coming up with an improved design.

Applicability

Every product development process should include competitive benchmarking.

Market benchmarking will be used most heavily during the opportunity development stage.

Technical benchmarking is used during the opportunity development stage to develop marginal and ideal values. It is also used during the concept development stage to provide reference concepts and help establish technical and performance models.

Design benchmarking is used most heavily in the subsystem engineering stage.

See also

Development Reference: Six Sigma (11.62), Quality Function Deployment (11.51), Value Engineering (11.69).

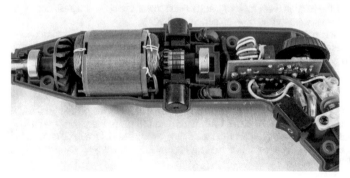

Competing products are thoroughly analyzed in design benchmarking.

Benchmarking

11.3 Bill of Materials

Track and manage each component in your design

Manufactured products are made up of various *specific* (not vague) components. The bill of materials is a table that lists each of those components.

The bill of materials, which is often called a BOM, generally includes information beyond the name of the part; for example, it can include the material of each part, the cost of each part, the part number, approved vendors, and more. It is organized in such a way that it makes it clear which components comprise each subassembly. It is a revision-controlled document that — when complete — is an extremely valuable part of a product's design information.

How to do it

1. Create a hierarchical table listing the system, subsystems, and components
2. Give each component a BOM item number for easy reference during discussion (typically the numbered rows of the table)
3. Give each subsystem and component a name and a description
4. List the approved manufacture for components purchased off-the-shelf
5. Give each component a part number; consider using a manufacture's part number for off-the-shelf components
6. Put the BOM under revision control
7. Update the BOM when new information becomes available

How to use it as a product development tool

1. Start a bill of materials as soon as your product concept has been selected
2. Include as much pertinent information as you can in the bill of materials when it is first created. Accept the fact that not all information will be known at first.
3. As the design evolves and more information becomes available, add it to the bill of materials and change the revision.
4. Add product development information, such as person responsible for the design of the component, whether or not a finite element analysis (FEA) is needed for the component, and so on.
5. Revise the BOM so that it always reflects the current state of the design.
6. Use the BOM in team meetings and design reviews to show progress and indicate where further development is needed.
7. Work to achieve a complete BOM, one that includes every specific component needed to manufacture the design. This will even include more obscure items such as labels, and paint as bill of material items.

Part lists in CAD assembly drawings

Engineering assembly drawings often include a table that lists the parts in that CAD assembly. In most cases, however, CAD models don't include every component in the product such as adhesive, components on a circuit board, labels, paint, wires, solder, and more. Because of this, these parts lists are not nearly as valuable as complete BOMs.

See also

Decomposition (11.14) and Drawings (11.23).

Bill of Materials shown is
for this assembly

Bill of Materials

Bill of Materials
Product: Drill
BOM Revision: C

Sub-Assembly: Cantilever assembly (Upper Structure, Boom Arm)
Revision Note: N. Toone, added finish and logo, 7-Mar-11

Item	Name	Description	Qty	Unit	Make/Buy	Drawing #	Revision	Part #	Approved Vendor	Unit Price	Extended Price
1	Cantilever assembly	Assembly, upper structure, boom arm	1	count	M	12200	B	12200	--	--	--
2	Cantilever beam	ASTM A-500 Grade C Steel Tube	1	count	M	13100	A	13100	--	16.14	16.14
3	Upright sleeve	ASTM A-500 Grade C Steel Tube	2	count	M	13200	B	13200	--	3.41	6.82
4	Horizontal hook member	1.25" OD steel rod	2	count	M	13300	A	13300	--	2.48	4.96
5	Vertical hook member	1.25" OD steel rod	2	count	M	13310	A	13310	--	2.10	4.20
6	Winch	3500 lb. capacity, automatic brake	1	count	B	13320	B	3196T63	McMaster-Carr	223.51	223.51
7	Winch cable, with hook	1/4" diameter, 25 ft. long	1	count	B	13321	--	3308T22	McMaster-Carr	52.97	52.97
9	Bolt, winch	M8 hex bolt, 30 mm, class 10.9, steel	3	count	B	13410	--	90854A170	McMaster-Carr	0.38	1.14
10	Nut, winch	M8 steel nut	3	count	B	13420	--	90592A022	McMaster-Carr	0.05	0.14
11	Washer, winch	M8 steel washer	3	count	B	13430	--	91455A130	McMaster-Carr	0.06	0.18
12	Pulley	4.25" OD, 3000 lb., steel, brass bushing	2	count	B	14000	--	31695T25	McMaster-Carr	58.15	116.30
13	Bolt, pulley	M12, 150 mm, Gr 8.8, st hex, DIN 931	2	count	B	14100	--	91280A749	McMaster-Carr	2.51	5.02
14	Nut, pulley	M12 steel nut, zinc coated, heavy duty	2	count	B	14201	--	90725A725	McMaster-Carr	0.55	1.10
15	Washer, pulley	M12 steel washer	2	count	B	14202	--	91166A290	McMaster-Carr	0.07	0.13
16	Spacer, pulley	1.75" length, 0.5" ID, LDPE	4	count	B	14200	--	92825A342	McMaster-Carr	1.10	4.39
17	Weld	See weldment drawing no 12300, Rev A	1.8	linear ft	M	--	--	--	--	20.00	36.00
18	Finish, powder coating	Blue (P: 300C), Polyester, 120 microns	1	count	B	--	--	--	Creer Metal Works	200.00	200.00
19	Logo	Vinyl cutout 3 inches by 3 inches	1	count	M	15100	A	15100	--	2.85	2.85
									TOTAL		675.85

Bill of Materials for an upper assembly prototype of the Human-Powered Well Drilling Machine.

11.4 Bio-Inspired Design

Apply solutions found in nature to your problem

Bio-inspired design (sometimes called *biomimicry* or *nature inspired design*) is a solution exploration strategy that investigates nature's solution to problems. The purpose of the strategy is to identify new solution principles from nature to inspire the development of analogous technical systems.

As an example, an innovative design for the leading edge of a wind turbine airfoil was inspired by a humpback whale. Humpback whales have protuberances or tubercles on the front edge of their fins. Studies show that these tubercles delay the stall angle of the fin by about 40%, increase lift, and decrease drag (Miklosovic et al., 2004). This allows the 40-50 foot whales to maneuver quickly to catch their prey (small fish).

This idea was applied to wind turbine airfoils. Instead of the traditional smooth blade, the humpback whale fin tubercles were imitated on the leading edge of the turbine blade and wind-tunnel tests have shown that these tubercles allow the blades to perform significantly better.

How to do it

When trying to solve an existing problem:

1. Search for similar problems in nature.
2. Evaluate nature's method for solving the problem.
3. Discover the underlying principle that makes nature's solution possible.
4. Determine if the same or a similar principle can be embodied in a product solution.

When trying to be inspired to create new products or technology:

1. Do an observational study, where the subject is nature.
2. Observe differences between the natural and man-made worlds.
3. Ask questions that promote thought: what makes a plant stand up? How does a tree pump water to its highest branches? How does a bee manage to fly?
4. Discover the underlying principle that makes nature's solution possible.
5. Determine if the same or a similar principle can be embodied in a product solution.

See also

Observational Studies (11.40); Janine Benyus' TED talks "Biomimicry in Action" and "12 Sustainable Design Ideas from Nature"; Delft University of Technology's "Bio-Inspired Designs" OpenCourseWare with lectures and readings.

Bio-inspired design of wind turbines. (a) Protuberances or tubercles on the leading edge of humpback whales. (b) Innovative turbine blade leading edge inspired by the humpback whale.

11.5 Brainstorming

Unlock the creativity of a group to find design solutions

Brainstorming is a group-based solution exploration method that builds on collective group knowledge and synergy to generate numerous candidate solutions to problems.

Brainstorming is a useful tool during any stage of product development. It can be used to generate ideas regarding requirements, ideas regarding solutions, ideas regarding potential failure modes, and any other thing that would benefit from group exploration.

How to do it

A successful brainstorming session is most likely to be achieved when it has been prepared for. To prepare for a brainstorming session, the person leading the effort should articulate the problem some amount of time (minimum 1 hour) before the session begins. This allows (i) the participants to familiarize themselves with the problem and the issues at hand, and (ii) it allows the person leading the effort to think clearly about who (meaning what expertise and background) should join the session.

The person leading the effort should prepare materials that will help the team use the session time effectively. This may mean that the leader has supplied paper, pens, and other props. It may mean that the leader has arranged inspirational solutions to be brought to the session. For example, the leader may bring industry leading competitive products, or other technological solutions that might inspire the session.

The leader should lead the session, meaning he or she should make sure there is a clear goal and that it is being met as much as possible. The leader should try to complete the session in less than one hour.

General brainstorming guidelines

Osborn's rules for brainstorming (Osborn, 1963):

1. Focus on quantity: The quality of ideas is not as important as generating as many ideas as possible for discussion. Having a larger group of ideas to talk about and choose from will lead to a higher occurrence of good ideas, which leads to a better final product.
2. Withhold criticism: Remember that concept generation and concept selection are two separate activities. Resist the impulse to criticize the infeasibility of suggested design candidates during the brainstorming session. Following this rule creates a comfortable atmosphere so team members share more ideas.
3. Welcome unusual ideas: Sharing wild ideas can help the group think in a new direction and generate ideas they otherwise could not. This will ensure that a broad range of solutions are explored. Even if the initial solution is not a good idea, it may lead to something that is.
4. Combine and improve ideas: Building on the ideas of others and combining different ideas allow the team to explore a larger design space so they can converge on a better solution.

See also

Development Reference: Bio-Inspired Design (11.4), Method 635 (11.35).

Further Reading

Stanford d.school's "Rules for Brainstorming," available at dschool.stanford.edu/blog/2009/10/12/rules-for-brainstorming

"Brainstorm Rules," available at dschool.stanford.edu/wp-content/themes/dschool/method-cards/brainstorm-rules.pdf

The Mind Tools website, www.mindtools.com/brainstm.html

Brainstorming.

11.6 CAD Modeling

Create virtual 3-D representations of your product

CAD modeling is used to create geometric representations of the product. Engineering drawings are typically created from CAD models, but CAD models can be effectively used far beyond the creation of engineering drawings. Effective use of CAD modeling can facilitate a wide range of activities and results in product and process development.

Key methods

3-D CAD solid modeling is generally recognized as preferable to 2-D drawing.

For high-quality surfaces, surface modeling is generally integrated with solid modeling systems.

Systems that support parametric modeling are preferred to those that do not.

Models should be defined in parametric form for at least two reasons. First, defining models parametrically identifies important design parameters. Second, models that have been defined parametrically are more robust and easier to change.

Systems that support parametric assembly of parts are preferred. In such systems, design based on a skeleton assembly can identify important assembly parameters and facilitate assigning modeling work to different team members.

CAD modeling can drive the design forward by forcing the designer to pick values for unknown parameters. This can be both good and bad. It is good, because it moves the design forward. It can be bad, because values can be chosen without appropriate consideration. As CAD models progress, be sure to identify Vital and Mundane parameters and ensure that Vital parameters are appropriately handled.

Engineering drawings that are automatically created from CAD systems are unlikely to be complete and high quality. Be sure to allow time for creating excellent drawings from the CAD models.

Drawings should always be checked by someone other than the creator of the drawing. When checking drawings, do not give the creator of the drawing the benefit of the doubt. Assume that there are mistakes, and work hard to find them.

Applicability

CAD modeling is generally used in the subsystem engineering and system refinement stages of development. However, CAD models can be very helpful in concept development, particularly as the concepts get more refined.

Most studies that have been made indicate that too heavy a reliance on CAD modeling early in the concept development stage can have a negative effect on the quantity and variety of concepts developed.

See also

Development Reference: Finite Element Modeling (11.30).

Further Reading

Veisz D, Namouz EZ, Joshi S, Summers JD, CAD vs. Sketching: An Exploratory Case Study, http://www.clemson.edu/ces/cedar/images/5/59/2013-veisz-sketching.pdf

CAD Modeling

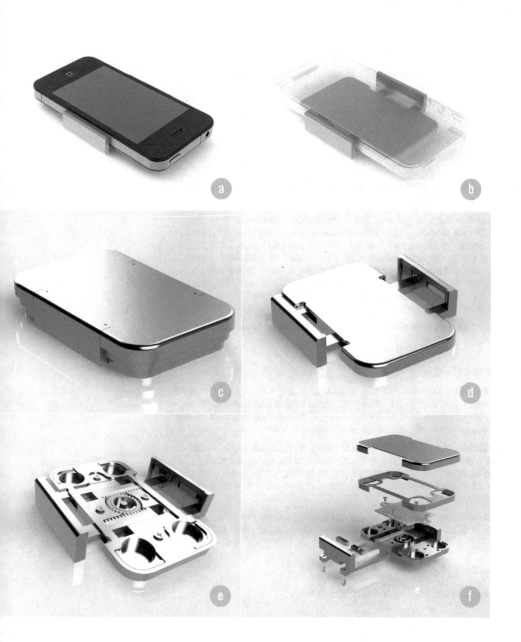

CAD Modeling

Renderings from CAD models can be used to (a,b) visualize things that would be difficult with physical prototypes, (c) show photo-realistic esthetics, (d-e) show motion, and (f) show relationship between parts in a product.

11.7 Catalog Search

Take advantage of existing components

Catalogs and manufacturers' technical information are sources of existing information that can be invaluable to a product development project. Catalogs are available both online and in paper copies. Paper catalogs are often easier to browse through than electronic catalogs, while electronic catalogs are often easier to search for specific items.

Manufacturer catalogs

Catalogs from specific manufacturers, such as Timken for bearings or Suspa for gas springs, generally contain the most specific information and are very helpful for details when a concept has been chosen.

It is very common that catalogs or design guides that are available from manufacturers will provide specific information about how to design your product to make effective use of the manufacturer's product. For example, the Timken Engineering Manual[1] provides a bearing selection procedure, installation methods, shaft and housing fit guidelines, and several pages for determining the loads on the bearings. This information is an invaluable resource that enables you to properly select bearings for your application and properly design your product to take advantage of the bearings.

Design guides from manufacturers help ensure high-quality designs using their components.

Distributor catalogs

Catalogs from distributors, such as MSC or McMaster Carr, contain a wide range of products and may be useful for browsing just to see what is available. Distributor catalogs tend to have much less technical information than manufacturers catalogs. However, they have a much broader range of available components.

When using distributor catalogs, it can be tempting to exclusively use a favorite distributor. When you know exactly what you want, and the distributor has it, using the favorite distributor is wise. However, if you are looking for ideas, it is generally best to look at multiple distributor catalogs, because each will have a different range of available components.

Catalogs for consumer, rather than industrial, products are often very helpful in finding ideas that can be used as the basis for a design solution, even though the product itself is unlikely to be used.

See also

Development Reference: Observational Studies (11.40).

---

[1] Available at http://www.timken.com/en-US/products/Pages/Catalogs.aspx

TIMKEN ENGINEERING MANUAL

The Timken Engineering Manual is a design guide for the selection and application of various rolling element bearings.

The MSC catalog contains thousands of components that can be useful in product development.

11.8 Codes and Standards

Apply the combined experience of the profession

Many products will be governed by relevant codes and standards. In such cases, the standards describe performance measures that must be met in order to sell the product. The standards also often prescribe test methods for measuring the required performance. Using codes and standards helps ensure your product will be successful, and allows you to build on the expertise of the standard writers, rather than having to create all of that knowledge yourself.

Codes

Codes are generally regulatory in nature. Codes describe characteristics a product must have if it is permitted to be sold. Code compliance serves as a constraint for a product. If a product does not meet the relevant codes, it doesn't matter how well it performs in other areas. It will be unacceptable.

Codes can be developed by professional societies, by trade organizations, or by government bodies. Regardless of how they are created, codes become applicable when they are accepted as mandatory by a government or other regulatory body.

Standards

Standards are developed by governments or trade associations to ensure compatibility and interoperability of parts between different manufacturers. When parts meet standards, they will work with other parts that meet the same standards.

Unlike codes, in many cases standard compliance is voluntary. A team can develop a product that uses non-standard threads, for example. However, the market will generally require that products comply with most existing commercial standards. Thus, understanding and complying with standards is generally good design practice.

Obtaining codes and standards

Codes and standards are generally available for purchase from the body that creates the code or standard. These fees provide the resources necessary to update and maintain the standards.

Determining which standards apply

Companies who have experience in developing specific products generally have a good understanding of the relevant codes and standards. For designers who are new to the development of a specific kind of product, codes and standards can often be identified by reviewing technical literature of related products and by the Delphi method.

See also

Development Reference: Delphi Method (11.15).

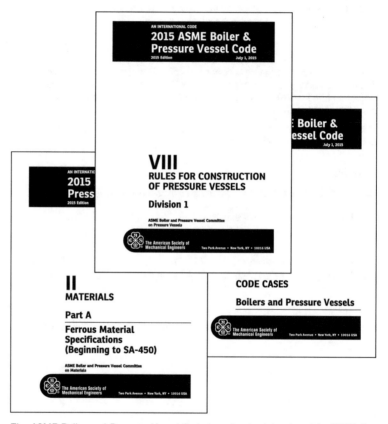

The ASME Boiler and Pressure Vessel Code is a standard developed by ASME that has been adopted as a code in multiple nations.

11.9 Concept Classification Tree

Organize your ideas to better cover the design space

A concept classification tree graphically presents the different branches explored when generating alternative solutions or concepts. In addition to providing a measure of variety, the classification tree may provide additional ideas by highlighting branches that are not sufficiently explored.

How to do it

The concept classification tree is constructed after generating a variety of concepts. The first step is to group similar generated concepts together. The second step is to simply capture the structure of the set of generated concepts in branching tree format as shown in the figure.

This can be done for all decomposed parts of the problem together, or for just one branch of the decomposition, as shown.

Benefits

Ulrich and Eppinger (2012) suggest four benefits of concept classification trees:

1. Pruning of less-promising branches.
2. Identifying different approaches to solving the problem (each branch is an independent approach).
3. Exposing insufficiently explored branches.
4. Refining the decomposition of particular branches.

See also

Development Reference: Decomposition (11.14), Recombination Table (11.53).

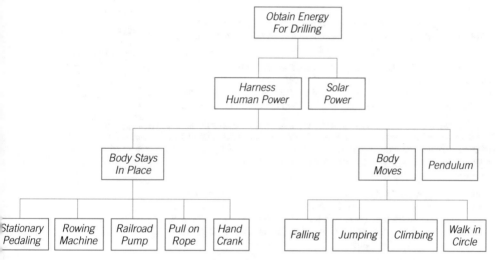

A concept classification tree used by the human-powered drill team (see Appendix B) to explore the "harness energy" subfunction. Note that although the team had as their primary focus the creation of a human-powered drill, during concept generation they expanded their thinking to consider solar power as a potential energy source. Perhaps their solution set would have been even better had they considered wind and rain as potential energy sources as well. By placing the concepts in a tree that shows logical relationships, we can see areas that could profitably be explored more fully.

Concept Classification Tree

CD

SSE

11.10 Controlled Convergence

Develop a concept that is demonstrably superior

Controlled Convergence is a formal method for creating a strong concept, using matrix methods to drive a significant understanding of market requirements and the development of concepts that are demonstrably superior at meeting those requirements.

How to do it

Create a convergence matrix with market requirements listed on the left, and potential solution concepts listed on the top, with a reference or datum concept on the left-hand side.

Next, rate the performance of each concept relative to the market requirements, working one requirement at a time (row by row). Concepts are given a "+" if they are better than the datum for that requirement, a "-" if they are worse than the datum, and an "S" if they are the same as the datum or the team cannot agree on a rating.

During the rating stage, each team member seeks to understand the reasons for the ratings. This is not a time for members to agree with one another to avoid conflict. It is a time to think critically about each of the concepts and requirements, and develop rationally based ratings. The understanding that comes from thorough evaluation is even more important than the actual rating.

When the ratings are complete, try to eliminate dominated concepts and combine complementary concepts. Concept B is said to dominate concept C if concept B is better than concept C in at least one rating, and has no ratings worse than concept B. Thus, there is nothing about concept C that is better than concept B. Dominated concepts can be removed from the matrix.

New concepts are created based on combinations of complementary concepts. Concepts D and E are complementary if concept E is strong in areas where concept D is weak and concept D is strong in areas where concept E is weak. Find ways to combine complementary concepts to maximize the strengths and minimize the weaknesses. Combined concepts are added to the matrix and rated relative to the datum. Dominated concepts are removed again. This completes Phase I.

Phase II has the same activities as Phase I, but with two important differences. First, the datum for Phase II is chosen from among the strongest concepts identified in Phase I. Second, the concepts in Phase II are described at a higher level of detail, which permits finer resolution in the evaluations.

If no clear dominant concept has emerged after the completion of Phase II, a third phase is undertaken. Again, the datum is changed, and more detail is added to the remaining concepts. By the end of Phase III, a dominant concept will almost certainly be identified.

Applicability

Controlled convergence method should be used when a concept needs to be developed for one of the Vital Few decisions. It should not be used for mundane decisions, because it takes too much time.

See also

Development Reference: Scoring Matrix (11.59), Screening Matrix (11.60).

Further Reading

Controlled Convergence is thoroughly explained in Pugh, S (1991) Total Design: integrated methods for successful product engineering. Addison Wesley, Reading Massachusetts.

Frey D, Herderi P, Wijnia Y, Subrahmanian E, Katsikopoulos K, Clausing D (2009) The Pugh Controlled Convergence method: model-based evaluation and implications for design theory. Res Eng Des 20(March):41-58. Available online at http://hdl.handle.net/1721.1/49448. This paper uses simulation to demonstrate that controlled convergence achieves consistent outcomes.

Market Requirements (Whats)	Rota-Sludge (Bench-mark in Tanzania)	People Walking in Circle	Tugging with Rope	Stationay Pedaling	Rowing Machine	Railroad Pump	Climbing or Falling	Pendulum	Hand Crank	Solar Power
The energy harnessing device provides enough torque to the drill bit.	S	+	+	+	+	+	+	+	+	S
The energy harnessing device drills the boreholes quickly.	S	+	+	+	+	+	+	S	+	+
The energy harnessing device requires few people to operate.	S	−	−	−	−	+	−	+	+	+
The energy harnessing device is a simple device.	S	+	+	−	−	−	−	−	−	−
The energy harnessing device is simple to operate.	S	+	+	+	+	+	+	+	+	+
The energy harnessing device is manufacturable in Tanzania.	S	+	+	−	−	−	−	−	−	−
The energy harnessing device allows a person to be stationary during operation.	S	−	−	+	+	S	−	S	S	+
The energy harnessing device minimizees the strass on the body.	S	−	−	−	−	−	−	+	−	+
The energy harnessing device is small.	S	+	+	S	S	S	−	−	S	S
Sum +	0	6	6	4	4	4	3	4	4	5
Sum −	0	3	3	4	4	3	6	3	3	2
Sum S	9	0	0	1	1	2	0	2	2	2
Total Entries	9	9	9	9	9	9	9	9	9	9
Disposition		Keep	Combine	Keep	Keep Research	Combine	Discard	Discard	Discard	Discard

A first-phase Controlled Convergence Matrix. Market requirements are listed on the left. Potential solutions are listed on the top, with the datum concept listed at the left side. Each concept is rated the same, better, or worse than the datum concept. Although the number of ratings in each category is totaled for each concept, no overall score is assigned to a concept.

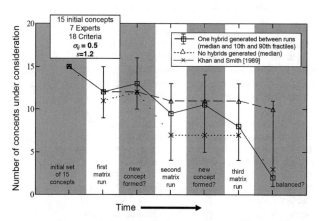

Convergence of concepts in controlled convergence. Three rating stages are demonstrated. In each stage, concepts that are dominated are eliminated. Between stages, new concepts are developed. After Frey et al.

11.11 Cost Estimation

Estimate the cost of mass producing your design

One factor that is often essential to product development is the cost associated with mass producing a design. In order to judge the profitability of an evolving design, it is helpful to estimate its cost during various stages of its development. To estimate the bill of material cost and other costs of a design, try one of the following.

How to estimate costs using prior knowledge about a similar product's costs

1. Make a deliberate choice about the level of accuracy (fidelity) needed for the cost estimate. Do the following steps consistently according to the chosen level of accuracy for all evaluations.
2. Select a sufficiently similar product or products for which you have cost data. This is easily done when developing a next generation product, where you can start with the previous generation's costs. Adjust the data according to inflation or other market dynamics to best reflect current costs.
3. Make intelligent estimates for as many of the costs shown in the opposing figure needed to match the chosen level of accuracy. Use the existing data as reference. For example, if the similar product uses 4 inches of 22 gage stranded core wire for $0.07, and the candidate concept requires 2 inches of the same wire, estimate $0.035 for the wire cost. Do this for each known part or subassembly of the concept.
4. Sum all of the costs to get a total cost estimate.
5. Consider multiplying the sum cost by a safety factor (e.g., 1.2, 1.5) to capture costs you choose not to model.
6. Update the cost estimates as new information becomes available.

How to estimate costs without prior knowledge about costs

1. Make a deliberate choice about the level of accuracy needed for the cost estimate. Higher fidelity is not always worth the time required. Do the following steps consistently according to the chosen level of accuracy for all evaluations
2. Collect pertinent general data for as many of the costs shown in the opposing figure needed to match the chosen level of accuracy. The general data collected would be cost per mass for raw materials, labor rates, and machine rates, for example.
3. Make intelligent estimates of cost by making the general data of the previous step specific to the candidate concept. For example, an estimate can be made for the raw material cost of a machined 6061 aluminum cylinder by multiplying the preprocessed mass of the cylinder by the cost per mass of 6061 aluminum. Likewise an estimate of the processing cost can be made by multiplying the processing time by the labor rate and machine rate. Do this for each part that will be processed from raw materials.
4. Estimate or acquire price quotes for purchased parts and subassemblies. Do this for appropriate quantities to reflect mass production quantities planned for.
5. Sum all of the costs to get a total cost estimate.
6. Consider multiplying the sum cost by a safety factor (e.g., 1.2, 1.5) to capture costs you choose not to model.
7. Update the cost estimates as new information becomes available.

Further Reading

Ulrich KT, Eppinger, SD (2012) Product Design and Development, 5th ed. McGraw-Hill, New York. pp. 256-267.

van Boeijen A, Dallhuizen J, Zijlstra J, van der Schoor R (2014) Delft Design Guide BIS Publishers. pp. 152-153.

For international and national labor rates visit the Bureau of Labor Statistics at http://www.bls.gov/

For raw materials, search the internet for commodity material costs. For example, search "cost of 6061 aluminum per kg", then use typical/average costs.

How to use cost estimates

During the product development process, the current cost of producing the design (by a production system) should be periodically estimated and compared to the cost targets described in Cost Targets (11.12) of the Development Reference. To be most useful the same level of accuracy should be used for both the cost targets and the cost estimates.

See also

Development Reference: Bill of Materials (11.3), Benchmarking (11.2), Cost Targets (11.12), and Financial Analysis (11.29).

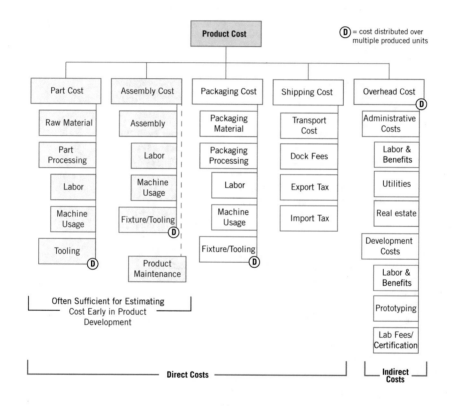

List of costs to be considered when estimating the cost of a product concept. Note that labor and machine usage are calculated by multiplying processing time by labor rate and machine rate, respectively. Product maintenance refers to the ongoing maintenance required to keep the product working desirably.

11.12 Cost Targets

Establish target bill of material cost based on product selling price

Product cost, selling price, and other financial characteristics of a design often drive decisions during the product development process. To that end it is valuable to establish a target bill of material (BOM) cost early in the development process. Having a target cost helps the team better understand how their design decisions affect the profitability of the product.

Basic concepts

When the consumer purchases the product in a store, the price they pay covers a lot more than just the cost of the parts. It also covers overhead and profits for each entity in the supply chain. For example, assume there are four entities at play: the manufacturer, the developer, the retailer, and the consumer. Assume you're the developer and have an interest in being financially profitable. To do so you'll need to consider the following:

BOM cost: This is the amount of money required to make the product including the sum cost of all the parts. Some companies also include labor/assembly costs in this amount (as a BOM line item).

Manufacturer's selling price: Assuming you will purchase the assembled product from a manufacturer, you'll need to pay the manufacturer more than the BOM cost. The added amount paid is known as the manufacturer's margin. The BOM cost plus the manufacturer's margin is the manufacturer's selling price to you.

Your selling price to retailer: Assuming you will distribute your product through a retailer, you will need to sell your product to the retailer for more than the manufacturer's price. The added amount charged to the retailer is known as your margin. The manufacturer's price plus your margin is the selling price to the retailer.

Selling price to consumer (end user): The selling price of a product is the amount of money the consumer pays to acquire the product from the retailer. The difference between the selling price to the customer and the selling price to the retailer is the retailer's margin.

Overhead: All of the entities at play have a certain amount of operating costs that are indirectly related to the product. For example, the developer will need to have an office with lights, computers, restrooms, and a staff of people to run the business. These costs can generally be grouped into one category called *overhead*. Each company's overhead is different. Nevertheless product development teams can use rough estimates to get a feel for the financial viability of a product by estimating overhead to be 50%. For example, if a manufacturer's selling price to you is $20, your overhead for that unit will be $10, or the overhead rate times the manufacturer's selling price.

Revenue: The revenue is income from the sale of goods. For example, if your selling price to a retailer is $40 and 10,000 units are sold, your revenue is $400,000.

Profit: The profit is the revenue (income) minus all of the costs (expenses). Generally the costs include the *cost of goods sold* (COGS), which is the manufacturer's price if you are the developer, plus overhead. For example, if you paid $20 per unit to the manufacturer and sold that same unit for $40 to a retailer, and your overhead was 50%, you would make $10 profit for every unit sold. If you sold 10,000 units, your profit would be $100,000.

Further Reading

van Boeijen A, Dallhuizen J, Zijlstra J, van der Schoor R (2014) Delft Design Guide BIS Publishers. pp. 152-153.

Pahl G, Beitz W (2007) Engineering Design, 3rd ed. Springer-Verlag, London. pp. 560-561.

For price theory, see Varian H (1992) Microeconomic Analysis, 3rd ed. WW Norton & Company, New York.

How to estimate a target BOM cost based on product selling price

- Determine a required (or estimated) selling price for the product you're designing. Price theory – not discussed here – is a complex topic ultimately used to determine selling prices. In the early stages of product development, however, price *estimates* can and should be made to guide the team. This is often done by a simple market assessment of similar products or of products having similar parts and complexity (see Benchmarking (11.2)).

- Divide the required (or estimated) selling price by approximately 8 to calculate a target BOM cost.

As a rule of thumb, 8 times the BOM cost is roughly what consumers pay when buying a product at a store. This is illustrated in the image provided here, and typically applies to small consumer goods. For larger purchases such as a car, the ratio is typically lower. When different ratios are known, the known ratios should be used.

How to estimate selling price starting with BOM cost

- Assemble a bill of materials (BOM), and calculate total BOM cost.

- Double the BOM cost to get an estimate of the manufacturer's selling price to you.

- Double the manufacturer's selling price to you to estimate your selling price to a retailer.

- Double your price to a retailer to estimate what the consumer will pay when buying the product off the shelves of a store.

If you've received a quote from a manufacturer for an assembled product, recognize that this quote has the BOM cost plus the manufacturing/assembly costs plus the manufacturing margin in it. In such cases an estimated selling price to a consumer is made by multiplying the manufacturer's quote by approximately 4.

See also

Development Reference: Bill of Materials (11.3), Benchmarking (11.2), Cost Estimation (11.11), and Financial Analysis (11.29).

Cost Targets

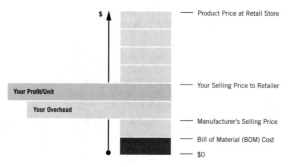

An illustration of the bill of material costs (dark gray) versus the product price at a retail store. This is a safe estimation for consumer products.

11.13 Critical Path Analysis

Identify the most critical parts of your project plan

Further Reading

Ulrich KT, Eppinger SD
(2012) Product design and
development, 5th ed.
McGraw Hill, New York. pp.
384-385.

In any design activity map, there is a critical path that limits the minimum time required to complete the map. An important characteristic of the critical path is that if any activity on it is delayed, it delays the completion of the entire map. Activities not on the critical path have a window within which they can be completed; any delay within that window does not delay the completion of the entire map. To that end, understanding, tracking, and managing the critical path are essential to completing the map on time.

Critical path analysis is often used when establishing a project schedule. As shown below, it can be used to establish (i) the necessary start times for each critical path activity, and (ii) the earliest and latest possible start times for non-critical path activities.

Multiple commercially available software packages can be used to identify the critical path. The backbone of these packages is simple to understand; for many small-scale projects critical path analysis can be carried out manually as fast as it might take to learn a new software package.

How to do it

1. Create an activity map (see Section 2.3).
2. Establish a duration for each activity in the activity map.
3. Using activity map logic[1] determine the earliest possible start time for each activity (do this by following the map in a forward direction).
4. Using activity map logic, determine the latest possible start time for each activity (do this by following the map in a backward direction).
5. Subtract the earliest possible start time from the latest possible start time to determine the size of the starting window (this window is often called the float time, in other literature).
6. Identify the critical path as the set of activities with starting windows of zero duration.

Simple example

Consider the activity map shown in the figure. Notice that steps 1 and 2 above are complete, meaning that each design activity (alpha designated arrows) in the network has a duration associated with it (number appearing next to it, e.g., T = 3 wks).

To determine the earliest possible start time we begin with the first activity (A). Clearly the earliest possible time to start activity A is beginning the network or time = 0. We then go on to the next activity (B), where based on the activity map logic that activity A must precede activity B, it can be seen that the earliest possible start time for activity B is time = 3. This process is continued for each activity.

Notice that because of the interdependency between Outcomes 3 and 4, the earliest possible time to start activity G is time = 20. This is because G cannot begin until activity F and activity C and D are complete. Of those three activities, activity C has the longest duration at 15 weeks; adding those 15 weeks to the three weeks of activity A and two weeks of activity B tells us the earliest possible start time for activity G.

[1]See Figures 2.7 and 2.8

To determine the latest possible start time, we begin with the fact that the network is expected to be completed in 32 weeks. We then start at the end of the network and work backwards; activity I has a duration of 4 weeks and must be completed by the end of week 32. Therefore the latest possible time to start activity I is week 32 minus 4 weeks, or at week 28. Further, the network shows that activity H must precede activity I, and that activity H has a duration of 8 weeks. Therefore the latest possible time to start activity H is the latest possible time to start activity I minus the duration of activity H, or at week 20.

A table showing the earliest possible start times and the latest possible start times is a convenient way to manage the data. With these times in hand, the starting window size can be calculated as shown.

The activities with a starting window size of zero are on the critical path.

See also

Development Reference: Concept Classification Tree (11.9), Recombination Table (11.53).

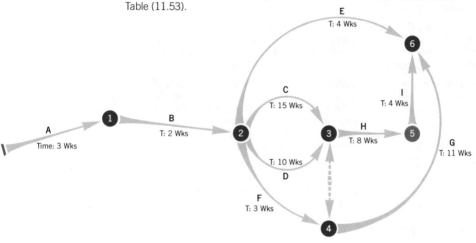

Activity	Earliest Start Time (ES)	Latest Start Time (LS)	Starting Window Size = (LS-ES)
A	0	0	0
B	3	3	0
C	5	5	0
D	5	10	5
E	5	28	23
F	5	18	13
G	20	21	1
H	20	20	0
I	28	28	0

Activity map for critical path analysis. The critical path consists of activities A, B, C, H, and I.

CD

11.14 Decomposition

Solve difficult problems one piece at a time

Decomposition – or the act of breaking something down into simpler constituents – is a universal design tool, used in nearly every design project. We can use decomposition to break down a complex project, a complex product, and complex functions. Such breakdowns make it possible for the product development team to attack smaller more manageable design sub-problems.

The need for decomposition varies but includes the fact that different expertise may be required to develop each subsystem, or that the project schedule is tight and requires parallel development in order to complete the project on time.

How to do it

Structural decomposition

The easiest strategy to understand is Structural Decomposition. Structural decomposition is the act of breaking down a product into its structural or physical parts. For example, consider a bicycle. It can be structurally broken down into (i) a welded frame, (ii) wheels, (iii) seat, (iv) handlebars, (v) fork, and (vi) drive train.

Structural decomposition is a valuable activity when a basic structure/concept already exists. If developing a new bicycle for a traditional bicycle market, structural decomposition is a good place to start because the traditional market will expect wheels, seats, handlebars, and so on in a new bicycle. With such a decomposition it is easier for the development team to decide which parts it will purchase off-the-shelf (drive train, for example) and which part it will develop a new design for (welded frame, for example).

Structural decomposition also facilitates project management, as part numbering, document control, development assignments, and so on are often made according to the basic physical parts of the product.

Functional decomposition

A more abstract (less tied to a specific structure or concept) decomposition strategy is functional decomposition. Functional decomposition is the act of dividing the top-level function into simpler subfunctions.

Again considering the bicycle, the top-level function is human transportation. To achieve such function, subfunctions include: (i) structural (hold everything together), (ii) contact ground, (iii) support rider, (iv) rider control, (v) power vehicle, and (vi) esthetics.

Functional decomposition is a valuable activity when there is no established concept or preconceived structure. It's a powerful way to abstract to the essence the problem being solved. When this decomposition is made explicit, it can greatly facilitate the development process as it provides significant design freedom because it is not tied to a specific structure. Notice, for example, that any human-controlled ground vehicle could result from a development project based on this functional decomposition – not merely a bicycle.

See also

Development Reference: Concept Classification Tree (11.9), Recombination Table (11.53).

Further Reading

Ulrich KT, Eppinger SD (2012) Product design and development, 5th ed. McGraw Hill, New York. pp. 121-123.

Herrmann JW (2004) Decomposition in product development, TSR 2004-6. Institue for Systems Research, University of Maryland.

Decomposition

Structural Decomposition

Welded Frame
Wheels (hub, spokes, rim, tube, tire)
Seat
Handle Bars
Drive Train
Fork

Functional Decomposition

Structure
Contact Ground
Support Rider
Rider Control
Power Vehicle
Aesthetics

Structural and functional decomposition for bicycle.

Delphi Method

11.15 Delphi Method

Obtain the help of experts on your product development challenges

The Delphi method or Delphi technique is a method for obtaining information relevant to the design from experts in related fields.

There is an informal Delphi method and a formal Delphi method.

Informal Delphi method

The informal Delphi method is based on interpersonal networking. The person looking for existing information finds a few people who might have ideas about where such information might be found. Each of these people is contacted, and asked to provide guidance about finding the desired information. At the end of the interview, the person is also asked for the contact information of other people who might be able to contribute to the search.

The informal Delphi method is helpful because you are drawing on the expertise of others without making large demands on their time. Using only a few minutes from any one person, the team is able to get access to a broad collection of expertise. The results from the informal Delphi technique are often used as a starting point for internet or catalog searches.

Formal Delphi method

The formal Delphi method uses a panel of 15 to 50 experts, with from 3 to 5 rounds of responses. The initial round consists of an open-ended question asking the experts to express their judgment about where information might be found or what might be a good concept for achieving a desired design outcome. The responses are then summarized, and the summary sent to the experts for a second round, where the experts are asked to rate and/or rank the responses. In the third round, the rated responses are returned to the experts, who are then asked to explain any disagreements they may have with the composite rating. The process potentially continues for a fourth and fifth round.

Hopefully, as the rounds continue, a consensus develops among the experts, thus providing stronger evidence of the goodness of the ratings.

An advantage to the formal Delphi method is that significant expert input is obtained with a relatively small amount of effort from any one expert. A disadvantage to the formal Delphi method is that several days or weeks may elapse between rounds, so that it takes a relatively long time to complete.

See also

Development Reference: Brainstorming (11.5), Observational Studies (11.40).

Further Reading

Hsu, CC, Sandford BA (2007), The Delphi Technique: Making Sense Of Consensus, Practical Assessment, Research & Evaluation 12(10). Available online at http://pareonline.net/pdf/v12n10.pdf.

The Delphi method uses experts to obtain information helpful to the development of the product.

11.16 Design for Assembly

Make your product easier, less expensive, and better to assemble

Design for assembly is the process of considering and improving the design so that it can be used to assemble the finished product more efficiently while still meeting the non-assembly related design requirements. Generally, the earlier design for assembly influences the design, the smoother and more compliant to the development budget and milestones, the production ramp-up[1] will be.

Principles of design for assembly

The following principles[2] guide the design for assembly activities:

- Minimize the total number of parts: Essential parts are those that (i) have motion relative to other essential parts, (ii) are necessarily made from a different material than its surroundings, and (iii) would not be possible to assemble unless it's separate. All other parts are candidates for elimination.
- Minimize assembly surfaces: These surfaces are costly to prepare. For example, it is expensive to machine parts on a cast engine block to prepare the locations where it mates with other engine components. The design should be improved to minimize assembly surfaces or their need for preparation.
- Avoid separate fasteners: Using integral fasteners (such as snap fits and heat stakes) reduces the total number of parts. If separate fasteners must be used, use a minimal number of unique fastener types.
- Minimize assembly directions: The product should be developed so that it can be assembled completely from one assembly direction. For example, a main component (case) can be placed on the assembly table and all parts assembled to it from a downward direction perpendicular to the table.
- Maximize lead-in assembly: The generous use of tapers, chamfers, radii allows parts that are not perfectly aligned to assemble together.
- Minimize handling in assembly: The basic handling operations during assembly are: picking up, orienting, and joining parts. The design should be refined so that the number of times and length of time required for these operations are minimized.

It's worth noting that to facilitate assembly, some design features will be added that are not needed for the product function. Because the products we design will not last forever, we must also consider these principles as they relate to disassembly.

See also

Development Reference: Design for Manufacturing (11.17).

Further Reading

Boothroyd G, (1994) Product design for manufacture and assembly. Comp Aided Des 26(7):505-520.

Boothroyd G, Dewhurst P, Knight WA (2010) Design for Manufacture and Assembly 3rd. ed. CRC Press

[1]This concept is discussed in Section 7.2.

[2]Derived from (Boothroyd et al., 2010) and (Dieter, 2000)

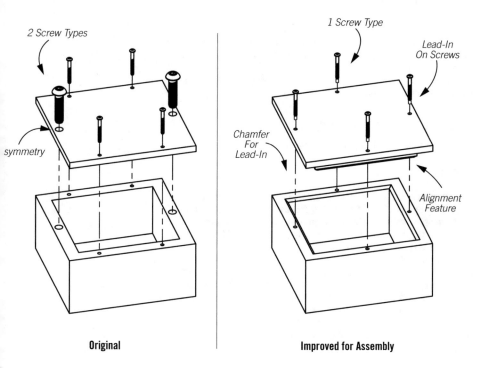

Original | **Improved for Assembly**

Example of improvements that result from the design for assembly guidelines.

11.17 Design for Manufacturing

Make a product better and less expensive at the same time

Design for manufacturing is the process of considering and improving the design so that it can be manufactured efficiently while still meeting the non-manufacturing related design requirements.

Seasoned designers start thinking about the manufacturing process very early in the development process, often in the concept development stage, and sometimes in the opportunity development stage. Generally, the product is more likely to be developed within budget and according to schedule when design for manufacturing activities is carried out early in the development process.

Principles for design for manufacturing

The following principles[1] guide the design for manufacturing activities:

General guidelines

- Become familiar with and plan for specific manufacturing processes: There are specific guidelines for specific manufacturing processes, such as injection molding, machining, or casting. The development team should choose an appropriate process and become familiar with the associated guidelines. A good place to start is (Boothroyd et al., 2010).
- Minimize total number of parts.
- Standardize components: Do this within the product itself and across product lines. Product family design (Jiao et al., 2007) is highly related to this guideline.
- Avoid secondary operations: Deburring, heat-treating, polishing, plating, and other secondary operations should be eliminated if not necessary.
- Avoid tolerances that are too tight: Tolerances set tighter than needed inflate the manufacturing cost. Tolerance analysis techniques (Chase and Parkinson, 1991) are designed to help designers specify tolerances with care.

Specific rules

- Specify component sources: The team should specify each component to be (i) off-the-shelf and from whom it should be purchased, or (ii) built in-house and from which revision of the design it should be built. Sufficient information should be provided to remove all ambiguity.
- Specify manufacturing process: The team should avoid ambiguous specification of process, such as "polished surface." Instead it should unambiguously specify a surface and say "This surface to be polished to 0.4 μm using an electrolytic grinding process."
- Establish appropriate dimensions: The team should dimension the part according to a minimal number of datums to avoid tolerance stack-up, and should dimension to physical locations on the part and not to points in space.
- Develop parts for minimal reorientation: Ideally the part would not have to be reoriented, or refixtured during the manufacturing process.
- Define suitable fits: The team should specify proper clearances to result in the desired fit. This should not be left to interpretation by the manufacturing entity.

See also

Development Reference: Design for Assembly (11.16).

Further Reading

Boothroyd G, (1994) Product design for manufacture and assembly. Comp Aided Des 26(7):505-520.

Boothroyd G, Dewhurst P, Knight WA (2010) Design for Manufacture and Assembly 3rd. ed. CRC Press

[1]Derived from (Boothroyd et al., 2010), (Pahl et al., 2007), and (Dieter, 2000)

Design for Manufacturing

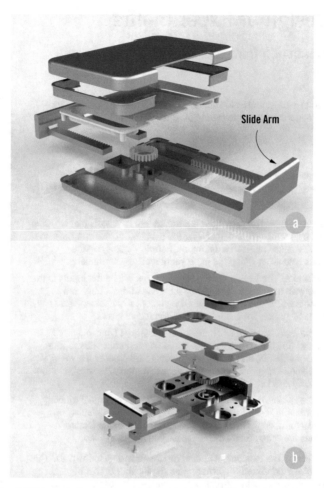

Slide Arm

Effects of DFM on design of a smartphone holder. (a) The design before serious DFM activities. (b) Design after DFM activities. The part was specifically designed to be machined. Consider the changes made to the slide arm. The original design is one piece, yet very difficult to machine. It was simplified by dividing it up into more machinable parts.

11.18 Design of Experiments

Get better information from your experiments by careful planning

Design of Experiments is used to develop statistical models that approximate the performance of a product based on the results of experiments or analysis.

How to do it

1. *Define the purpose:* Clearly identify the models you hope to obtain.
2. *Design the experiment:* Choose responses, factors, experiment type, and factor levels.

   a) Choose responses: The responses are the system behaviors that we hope to model by completing and analyzing the experiment. We can have one or more responses for each experiment. Formally specifying the responses before beginning the experiment focuses our thinking.

   b) Choose factors: The factors are the variables that will be the inputs to the model we wish to develop. The more factors we choose, the more experiments we will need to run. But if we ignore a critical factor that affects the response, the model will have significant errors.

   c) Choose the experiment type: There are at least three types of experiments that are commonly run: screening experiments, two-level factorial experiments, and multi-level factorial experiments. As we move from screening experiments to multi-level experiments, the cost and the information gained increase significantly.

   d) Choose the levels: At its core, experimental design consists of running experiments with factor levels chosen to maximize the availability of statistical information. The choice of levels for each of the factors determines the space over which the resulting model applies.

3. *Run the experiment:* The experiment should be run according to the plan. This will usually involve randomized run order and a significant number of experimental runs. Running the experiment as planned keeps the statistical information valid.

4. *Analyze the results:* Statistical analysis of the results produces the model that should achieve the purpose for the experiment. If the model is insufficient for its intended purpose, an improved experimental design is often created with fewer factors but more levels, and the process is repeated.

Applicability

DOE is often applied during the subsystem engineering stage to determine the best values for design parameters. It can also be used during any verification stage to determine the performance of the system or subsystem over a range of conditions.

While DOE requires numerous experimental runs, it has been developed to maximize the information obtained per run.

See also

Development Reference: Six Sigma (11.62), Quality Function Deployment (11.51), Uncertainty Analysis (11.68), Robust Design (11.57).

Further Reading

An excellent online resource for DOE techniques has been developed by the National Institute for Science and Technology, and is available at http://www.itl.nist.gov/div898/handbook/pmd/section3/pmd31.htm

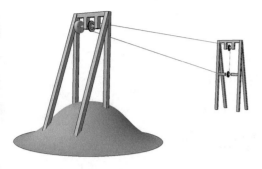

A Capstone team was asked to develop a power-generating zipline for students in rural African villages. The zipline would drive a generator that would charge a deep-cycle lead-acid battery. This battery would then be used to charge LED lanterns that the students could take home to illuminate their homework. The goal was to make the zipline fun, while at the same time extracting enough power to charge the lanterns.

An analytical model of the zipline speed and generated power was developed. This model, which included 6 simultaneous nonlinear differential equations, was only solvable numerically. It was necessary to choose the zipline height, length, pulley diameter, and generator gear ratio.

Run Number	Friction	Height	Radius	Stretch	Gear Ratio
1	0.1	1	0.1	0.4	1.75
2	0.3	1	0.1	0.4	1.25
3	0.1	3	0.1	0.4	1.25
4	0.3	3	0.1	0.4	1.75
5	0.1	1	0.3	0.4	1.25
6	0.3	1	0.3	0.4	1.75
7	0.1	3	0.3	0.4	1.75
8	0.3	3	0.3	0.4	1.25
9	0.1	1	0.1	0.6	1.25
10	0.3	1	0.1	0.6	1.75
11	0.1	3	0.1	0.6	1.75
12	0.3	3	0.1	0.6	1.25
13	0.1	1	0.3	0.6	1.75
14	0.3	1	0.3	0.6	1.25
15	0.1	3	0.3	0.6	1.25
16	0.3	3	0.3	0.6	1.75
17	0.3414	2	0.2	0.5	1.5
18	0.0586	2	0.2	0.5	1.5
19	0.2	3.414	0.2	0.5	1.5
20	0.2	0.586	0.2	0.5	1.5
21	0.2	2	0.3414	0.5	1.5
22	0.2	2	0.0586	0.5	1.5
23	0.2	2	0.2	0.6414	1.5
24	0.2	2	0.2	0.3586	1.5
25	0.2	2	0.2	0.5	1.8535
26	0.2	2	0.2	0.5	1.1465
27	0.2	2	0.2	0.5	1.5

A 27-run, 5-factor central composite experiment was developed. This experiment varied the friction, height difference, pulley diameter, rope stretch, and gear ratio. Each factor was evaluated at five different levels. The outputs for the experiment were maximum speed and average power output.

After analyzing the 27 experimental runs, a quadratic model was developed for both of the responses as a function of each of the factors. This figure shows a surface plot of the effects of pulley diameter and gear ratio as a function of pulley diameter and gear ratio, at a height of 2 meters, rope stretch of 0.5 meters, and a friction factor of 0.2.

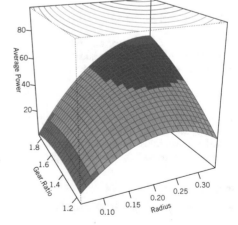

Slice at Friction = 0.2, Height = 2, Stretch = 0.5

11.19 Design Review

Assess your progress and get feedback and help from others

Design reviews are valuable events that take place multiple times during product development. They are typically held in the form of a short meeting. The purpose of a design review is to allow a reviewer to assess the desirability and transferability of the design, and to allow the product development team to receive feedback and help from others outside of the immediate team.

People who are not yet accustomed to design reviews often find the design review experience to be frustrating. To minimize frustrations, it is valuable to understand that a design review is a critique. Effective design reviews are centered on critiquing the design not the designer.

Design reviews exist in many forms, and different organizations have different traditions that define them, but they all exist to satisfy the principles listed here.

Principles that justify design reviews

- Small course corrections are easier and less costly to fix early than large course corrections later.
- People who are one step removed from the detailed work can see things those who are doing the detailed work cannot see.
- It's better for the team and the reviewers to find the problems than it is for the market to find the problems after the product is released.

Principles for success

- Seek to understand the objectives and goals of the design review before planning, leading, or participating in a design review.
- Develop a culture where design reviews are an open, two-way dialogue, with all participants trying to understand each other.
- Structure design reviews so that they identify errors or weaknesses in design. This is noticeably different than structuring the review to find only the strengths of the design.
- Focus the design review on advancing the design and facilitating development, more than trying to impress people or defend points of view.
- Make design reviews flexible so that they can be modified as needed for the benefit of design.
- Schedule regular, frequent, design reviews to facilitate many small course corrections.
- Plan for different kinds of design reviews, meaning reviews with supervisors, reviews with clients, reviews with end users, etc.
- Assign a member of the team to keep sharable notes during the review so the team can focus on the design and on understanding each other.

A starter outline

A basic design review outline is shown on the facing page.

See also

Development Reference: Basic Design Process (11.1), Project Objective Statement (11.49).

Design Review Outline

Quickly introduce and give context for the project. Keep in mind that the reviewers potentially have many projects/teams in their mind. Simply remind them about the project.

1. Project Objective Statement

2. Sharing of a recent great success

Plan for a brief celebration of success. Even though celebrating success is not the purpose of a design review, most team members will understandably want to do this. Planning a short amount of time for this can boost morale.

These are the core of the design review. They should occupy the largest amount of time and effort in the review. In 3, the desirability of the design is evaluated, with justification as to why design decisions were made. In 4, the roadblocks to greater desirability are discussed and possibly removed.

3. Review of key success measures and evidence on how well the team is doing on them

4. Discuss of the greatest challenges (in order of the most to least troubling)

5. Review of assignments resulting from design review

A brief moment at the end of the review should be spent clarifying any assignments and actions

A basic outline for a design review. When preparing for or leading a design review, modify the outline as necessary to achieve your review objectives.

Design Structure Matrix

11.20 Design Structure Matrix

Order your design activities for high quality and efficiency

The design structure matrix (DSM) is a matrix used to track the relationships between key parameters for a product. The DSM identifies sets of parameters that are independent, dependent, and coupled. It establishes an order in which parameters should be determined for a minimum amount of iteration, and thereby provides guidance for organizing activity maps.

Key methods

The traditional DSM is a matrix that lists all of the major activities used in a development project as labels for both the rows and columns of the matrix. The diagonal of the matrix is shaded. Marks are then placed at the intersection of a row and column when the activity in the row needs a result of the activity in the column. Thus, looking along a given row of the matrix, one can see all of the activities that must be completed in order to complete the activity in that row. Similarly, looking down a given column, one can see all the activities that need inputs from the activity in that column.

Interactions that are above the diagonal of the matrix indicate tasks that cannot be done in sequence, because an earlier task requires the output of a later task. This indicates places where a restructuring of activities may be beneficial.

Note that examples can be found in the literature where the meaning of rows and columns in the design structure matrix is reversed. If this is the case, interactions below the diagonal indicate feedback relationships.

Marks in cells of the matrix can be binary (simply the presence or absence of a mark), or they can be scaled, where different numbers are used to represent the strength of the interaction.

Once the matrix has been created, the tasks can be rearranged to minimize the number of above-diagonal interactions. This rearrangement will streamline the process and provide information in the order that it is needed.

Analysis of design structure matrices can be used to define subsystems, organize development teams, and optimize development processes. See Browning and Yassine *et al.* for more information on these uses.

Applicability

DSM is usually created during the opportunity development stage, as part of the project planning process.

See also

Sections 2.3 and 2.4 of this book.

Further Reading

An excellent review article on 30 years of DSM practice, with good insight into broad application of the DSM method, is Browning TR (2001) Applying the design structure matrix to system decomposition and integration problems: a review and new directions. IEEE T Eng Manage 48(3):292-306.

An example of how DSM can be applied to develop effective activity maps is reported in Yassine AA, Whitney DE, Lavine J, Zambito T (2000) Do-it-right-first-time (DRFT) approach to design structure matrix (DSM) restructuring. Proc DETC '00, DETC2000/DTM-14547.

A current book on DSM is Eppinger SD, Browning TR (2016) Design structure matrix methods and applications. MIT Press, Cambridge, MA.

Task	#	2	4	5	6	7	8	9	10	11	12	13	14	15	16	17	18	19	21	22	24	20	23	25	26	27
Select Material for All System Components	2									6	6							6		4						
Freeze Proportions and Selected Hardpoints	4			6	4								6													
Verify Hardpoints and Structural Joints	5		4		4					6																
Approve Master Sections	6		6	6			4																			
Generate Structural Requirements (Analytically)	7												3					4								
Develop Conceptual Design Strategy	8				3																					
Develop Structural CAD Model	9		6				4										4		4							
Verify Functional Performance (Analytically)	10	6					4																			
Develop Preliminary Design Intent CAD model	11				6				3				6		6	4		4			6					
Estimate Blank Size	12				4					4						4		4			3					
Estimate Efforts	13	6																								
Develop Initial Attachement Scheme	14				6					4						6		4			3					
Estimate Latch Loads	15				6					6																
Cheat Out Panel Surface	16	6								4					3			4								
Define Hinge Concept	17				4																3					
Get Preliminary MFG & Assembly Feasibility	18				4					4	6		6			4					3					
Perform Cost Analysis	19	6			6					4	4		3			4	4		4	4	3					
Theme Approval for Appearance	21				4												4				3					
Marketing Commits to Net Revenue	22																				4					
Approved Theme Refined for Craftsmanship Exec.	24	4																			4					
Perform Swing Study	20				4											4								6		
Program DVPs and FMEAs complete	23	4								4			3			4										
PDNO - Class 1A Surfaces To Engineering	25	4																			6	6	3		6	4
Conduct Cube Review and Get Surface Buyoff	26	4			4																3			6		4
Verify MFG & Assembly Feasibility	27				6					6	4		6			4	6				6	3		4	3	

Design structure matrix for design of automotive hood. After Yassine et al.

11.21 Dimensional Analysis

Understand how design parameters affect performance

Dimensional analysis is a method of determining a minimal set of parameters that will affect the performance of a design. It provides scaling factors that help compare different designs using the same principles. A result of dimensional analysis is a set of dimensionless groups that can be correlated with product performance across a wide range of conditions.

How to do it

The first step in dimensional analysis is to determine a complete set of n independent quantities that can affect the performance you are trying to model. The model is a model of n parameters:

$$P = \mathrm{f}(Q_1, Q_2, \ldots Q_n) \tag{11.1}$$

The quantities $Q_1 \ldots Q_n$ form a complete set if no other quantities determine the performance, and they are independent quantities if they can be independently adjusted.

Having obtained a complete set of independent quantities, we express the units of each quantity as a product of base units raised to powers. For example, mechanical problems have three base units: length, time, and mass.

$$[Q_i] = L^{l_i} M^{m_i} t^{t_i} \tag{11.2}$$

We next identify a complete, dimensionally independent subset of the parameters $Q_1 \ldots Q_k$, where $k \leq n$. A complete, dimensionally independent subset is one for which the dimensions of every quantity not in the subset can be expressed as a product of the dimensions of the quantities in the subset, and for which no quantity in the subset can have its dimensions expressed as a product of the dimensions of the other elements of the subset.

Once we have obtained the subset, we can express the dimensions of each element not in the subset as a product of dimensions of subset elements. We can then define a dimensionless form of the element as the ratio of the element and the product of the subset elements.

$$\Pi_i = \frac{Q_{k+i}}{Q_1^{N_{1i}} \ldots Q_k^{N_{ki}}} \tag{11.3}$$

We also express the dimension of the performance as a product of the dimensions of the subset, and then define a dimensionless value of the performance as the ratio of the performance and the product.

$$\Pi_0 = \frac{P}{Q_1^{N_{01}} \ldots Q_k^{N_{0k}}} \tag{11.4}$$

Having done so, we have then reduced the problem to a dimensionless problem.

$$\Pi_0 = \mathrm{f}(\Pi_1, \Pi_2, \ldots \Pi_{n-k}) \tag{11.5}$$

We have now reduced the problem from an n-dimensional problem to an $n - k$ dimensional problem. We have also established a number of dimensionless ratios Π_i that can be used in our experimentation or analysis of the problem.

See also

Section 3.5 of this book.

Further Reading

Sonin AA (2001) The physical basis of dimensional analysis 2nd ed. Handout for 2.25. Available at http://web.mit.edu/2.25/www/pdf/DA_unified.pdf

Wikipedia has an excellent article on dimensional analysis:
http://en.wikipedia.org/wiki/Dimensional_analysis

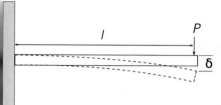

We shall apply dimensional analysis to understand the deflection δ of a cantilever beam of uniform section length l when a point load of P is applied. The beam material is assumed to be linear, isotropic, and homogeneous.
Based on the principles of mechanics, we expect the deflection to depend upon P, l, the modulus of elasticity of the beam E and the area moment of inerta of the beam cross section I.

The four independent quantities that may affect the deflection of the beam are composed of the three mechanical base units.

Quantity	Unit
P	MLT^{-2}
l	L
E	$ML^{-1}T^{-2}$
I	L^4

For a complete, dimensionally independent set of quantities we define l and E, since we can define all the remaining quantities by products of l and E, but we cannot define l and E by products of each other.

Quantity	Unit	Product	Π_i	Ratio
P	MLT^{-2}	$l^2 E$	Π_1	$P/l^2 E$
I	L^4	l^4	Π_2	I/l^4
δ	L	l	Π_0	δ/l

The products of l and E that are used to non-dimensionalize P, I, and δ are shown in the table.

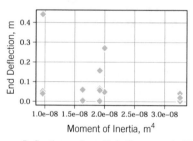

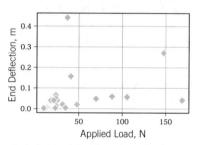

Deflection varies with both moment of inertia and applied load, but there is scatter in the data that indicates neither moment of inertia nor load totally predicts the deflection.

When the dimensionless quantities are plotted, the data all falls onto straight lines for each value of Π_1, indicating that the dimensionless parameters Π_1 and Π_2 are sufficient to predict the dimensionless displacement Π_0. The model has gone from four independent parameters to only two scaled parameters.

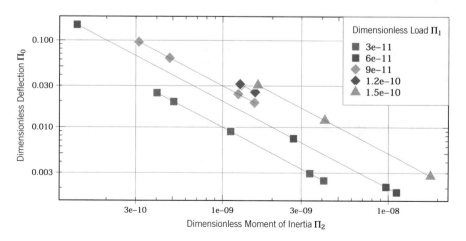

11.22 Drawing Checking

Make sure your design is clear, correct, and unambiguous

Before being approved for release, every engineering drawing must be thoroughly checked by an experienced designer or drawing checker. The purpose of checking the drawing is to make sure that there are no obvious mistakes on the drawing.

It is important to have the drawing check carried out by someone other than the person who created the drawing. The drawing creator is so familiar with the drawing that they are less likely to catch mistakes.

The drawing check will generally have at least two levels: Format Check and Design Check.

Format check

The format check is performed to ensure that all relevant standards related to the presentation of the drawing have been performed appropriately. Sample questions that might be asked during the format check include the following: Is the title block filled out completely? Has the appropriate drawing format been used? Are the proper number of views listed? Are the dimensions complete? Are the appropriate types of lines used in the drawing? Are the text heights correct?

The format check can be completed without knowing any details of the design intent or the function of the part. It simply evaluates whether the presentation meets the relevant standards.

The format check can be considered as roughly equivalent to spell checking a written document. It makes sure the elements are correct, but does not ensure that the document as a whole makes sense.

Design check

The design check is performed to ensure that the design as represented in the drawing is feasible and meets the design intent of the team. Therefore, the checker must understand the design requirements and the function of the part. Items such as surface finishes and tolerances are vital parts of the design check.

In general, the fundamental concept of the design should not be part of the design check. Fundamental concepts will have been previously evaluated in a design review. However, if the checker has concerns about the concept, it is better to raise them at this point rather than later.

The design check is similar to proofreading a written document. It makes sure that the content makes sense and meets the desired outcomes.

When both the format check and design check have been completed, the checker signs and dates the drawing in the appropriate location, indicating that the drawing has been checked.

See also

Development Reference: Drawings (11.23).

Further Reading

Hill engineering provides a recommended process for performing drawing checks: http://www.hillengineering.com/papers/How%20to%20Check%20a%20Drawing.pdf

Another procedure for checking drawings is available here: http://tinyurl.com/drawing-checking

Drawing Checking

Engineering Drawing Checklist
C. Mattson and C. Sorensen | R1.2

BYU
CAPSTONE

The following should be considered when checking drawings for completeness and appropriateness.

Format Check

✔	Description of Format Item to Check
	Title block – Is it completely filled out?
	Views – Do we have the proper number and kind to clearly illustrate the nature of the part or assembly?
	Dimensioning – Are standard practices followed?
	Dimensioning – Do we have tolerances and surface finishes specified?
	Dimensioning – If the part is plated, does the drawing specify dimensions are pre or post plating?
	Dimensioning – Are any parts overdimensioned?
	Units – Are the units clearly described as METRIC or INCH on the drawings?
	Symbols – Have industry-standard symbols been used in the drawings and diagrams?
	Materials – Is the material sufficiently specified so that it can be purchased from a general materials distributor? Or if necessary, from a specific distributor?
	Notes – Do they apply specifically to the drawing being checked?
	Notes – Are they complete?
	Reference – Is the drawing precise when referencing other documents?
	Approvals – Are the approver names and approval dates provided?
	Revisions – Is the current drawing revision accurately recorded with approvals?
	General – Can the drawing be misinterpreted?

Design Check

✔	Description of Design Item to Check
	Dimensioning – Do the dimensions accurately reflect the current product intent?
	Dimensioning – Do the dimensions make sense for the intended function?
	Dimensioning – Are the surface finishes appropriate for the part?
	Dimensioning – Are the tolerances appropriate for the part?
	Dimensioning – Is the precision of drawing dimensions appropriate?
	Dimensioning – Are the tolerances as loose as possible where they can be?
	Dimensioning – Are the datums appropriate for part function?
	Dimensioning – Are the dimensions relative to meaningful datums?
	Dimensioning – Do the critical dimensions and their tolerance result in a satisfactory tolerance stack up?
	Materials – For critical parts, are the chosen materials optimal for that part?
	Materials – For mundane parts, are the chosen materials sufficient for that part?
	General – Can the design be simplified?

A sample checklist for drawing checking.

11.23 Drawings

Create transferable design representations

Engineering or technical drawings are used to fully and clearly define the design of an engineered product. Drawings communicate all needed information from the designer of the product to the production system to ensure that the product will match the design intent in all important ways.

Types of drawings

Most drawings are pictorial drawings, meaning that they show the geometry of the part contained in the drawing. In addition to the geometry, drawings show the dimensions of the part, tolerances for the dimensions, material, and surface finish.

Assembly drawings are used to pictorially show an assembly and the components that make up the assembly. Components that make up an assembly may be either parts or subassemblies. Assembly drawings generally show only limited details about the components; the details are contained in part drawings. Assembly drawings have hidden lines removed.

Part drawings are used to convey the details of an individual part. Part drawings are required for every custom-designed part in a product. Pictorial drawings are not generally needed for purchased parts, but a specification sufficient to purchase the part is required.

Another common drawing is a schematic diagram. Schematic diagrams show logical relationships between components, rather than showing the geometry of the component. Schematic diagrams are common in electrical, pneumatic, and hydraulic systems.

Details of subsystem interfaces are often captured in interface control drawings. Interface control drawings describe geometrical, information, signal, and power interfaces between subsystems. They provide the standard which all subsystems must meet.

Some complex systems will also include block diagrams. Block diagrams show overall functionality of the system, with logical relationships between subsystems. Block diagrams are generally used when a schematic diagram of the whole system would be too complex to readily understand.

Non-graphical content

In addition to the graphical content of the drawing, there are important elements of technical drawings that are included by convention. These include the following:

- A title block, containing a drawing title, a drawing number, a part number, a revision number or letter, identifying information for the entity creating the drawing, measurement units for the drawing, default tolerances, general notes and specifications, and creator and approver information.
- A revision block, which lists all of the revisions made in the drawing from its initial release, including dates and a summary of the changes.
- A bill of materials, which lists all of the materials necessary to create the component in the drawing. The bill of materials is sometimes omitted from a part drawing.
- A list of notes, which provides textual information not included in the graphical drawing but necessary to properly create the component.

Drawings

• A drawing frame, which labels horizontal and vertical regions of the drawing with numbers and letters, thus facilitating discussion about the drawing.

Approval and revision control

In most cases, before a drawing can be released it must be approved by a number of different individuals. The title block should contain a place to record this approval. In most cases the title block lists the engineer, the drawing creator, the drawing checker, and approvals for various functions of the company such as production and testing.

The drawing should always be checked by someone other than the creator of the drawing.

When revisions to a drawing are required, an engineering change order should be prepared that describes the change. The change order should be approved by the entities who approved the original drawing. The drawing should be changed to be consistent with the change order. The revised drawing gets a new revision code and an updated revision history. The revised drawing should also be approved by the entities who approved the original drawing.

See also

Development Reference: Drawing Checking (11.22)

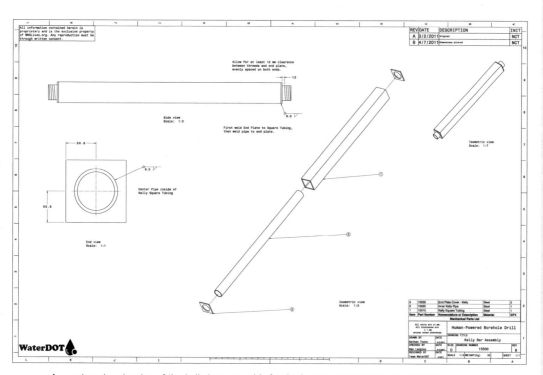

An engineering drawing of the kelly bar assembly for the human-powered water well drill described in the design case study.

11.24 Engineering Change Order (ECO)

Make design revisions in a calm and considered way

There comes a point in the evolution of a design where design changes need to be carefully coordinated between multiple people or groups. The engineering change order — or ECO — is a tool aimed at facilitating this. Specifically, an ECO is used to control the revision of documents and parts. It does this by making sure that the necessary stakeholders agree with how the change will be handled. It also serves as an official record of the change.

Parts of the ECO template

- The ECO is given a number for tracking/archival purposes, and a date when the ECO will take effect.
- The owner (creator) of the ECO and the product under consideration are listed, as well a brief description and the date the ECO was created.
- A *revise from* number to a *revise to* number is requested for specific documents and parts (as identified by their document and part number).
- A description of the change, the reason for it, and impact is also provided when the ECO is created.
- Disposition codes, as recommended by the ECO owner, indicate how the change will be handled once it goes into effect.
- Signatures of key stakeholders are obtained, indicating their approval of the change.

When to use an ECO

ECOs are generally used whenever a revised design has to be officially released by the core product development team to other entities. These other entities have a stake in the product, such as a manufacturer, a supplier, or an internal sales, purchasing, or management team. ECOs are also used in the subsystem engineering stage of development when large teams are involved and communication is not as fluid as it might be in a small team. In these cases, the ECO helps others on the team to know about key changes that may affect their work. This is particularly important when working on the interfaces between subsystems. A revision history in an engineering drawing can legitimately refer to an engineering change order (by its ECO number) as a way of capturing critical revision information in the drawing package. ECOs often include revised engineering drawings or component specification sheet as attachments that are referred to in the ECO.

See also

Bill of Materials (11.3), Drawings (11.23), Drawing Checking (11.22).

ENGINEERING CHANGE ORDER

ECO #: _____   **EFFECTIVE DATE:** _____

CO OWNER: _____	DATE: _____	
RODUCT: _____	DESC: _____	

NG CHANGE NOTICE (ECN) REQUIRED? ☐ YES ☐ NO
S CUSTOMER APPROVAL REQUIRED? ☐ YES ☐ NO   Pre-Production ECO: ☐ YES
S REGULATORY APPROVAL REQUIRED? ☐ YES ☐ NO   Release ECO: ☐ YES

OCUMENT #	FROM REV	TO REV	PART #	FROM REV	TO REV

REASON FOR CHANGE AND IMPACT:

DISPOSITION CODES

CLASS OF CHANGE		EXISTING PARTS		EFFECTIVITY
RECORD	1	USE	E	NEXT BUILD
2 WAY INTERCHANGEABLE	2	REWORK	F	NEXT PURCHASE
1 WAY INTERCHANGEABLE	3	RETURN TO VENDOR	G	PER SALES ORDER
NON-INTERCHANGEABLE	4	SCRAP	H	MANDATORY
			I	MANDATORY W/RETURN

ODE: DESCRIPTION OF CHANGE:
C1F   Implement when existing stock has been consumed ← Example Code

CO APPROVAL:

ECO Board Rep.: _____	☐ Engineering Rep.	_____	
Manufacturing Rep.: _____	☐ Quality Rep.	_____	
Procurement Rep.: _____	☐ Purchasing Rep	_____	
Sales Manager: _____	☐ Sustaining Manager	_____	
Supplier: _____	☐	_____	

c engineering change order template.

Ergonomics

11.25 Ergonomics

Create person-friendly designs

The interface between humans and technical products is called *product ergonomics*. When establishing product ergonomics, designers consider the characteristics, abilities, and needs of humans (Pahl et al., 2007) as well as the needs of the product.

A significant amount of research has been done in the area of workplace ergonomics and user interface (UI) design for software. Whether it be for workplace, software, or hardware, the human is the starting point. Specifically, three human issues should be considered (Cushman and Rosenberg, 1991). They are briefly discussed below.

Biomechanical issues

These issues relate to the human's body size and movement relative to the loads associated with using the product.

Henry Dreyfuss collected and cataloged anthropomorphic data for numerous subjects including the data shown on the opposite page. His full presentation of the data (Tilley and Associates, 2002) includes measurements of people of all ages, residential space considerations, vehicular accommodations, maintenance accessibility, and many other things. The data, which is for a US population, represents 98% of the population (the population between the 99th percentile and the 1st percentile). The 50th percentile means that half the population (assuming a normal distribution) is at or below this value, while the other half is at or above this value.

There are numerous examples of how this information can be used to influence a design. For example, it can influence control panel placement, grip size on power tools, size of office chairs, and much more.

Physiological issues

These issues relate to the human body's reaction to loads over time. All loads lead to some level of physical stress and fatigue. Clearly a load that can be exerted once may not be exertable continuously during a work day. Physiological measures that are often considered are, cardiovascular (heart rate or blood pressure), respiratory (respiration rate or oxygen consumption), and sensory (visual and hearing acuity or blink rate) (Sanders and McCormick, 1993).

Psychological issues

These issues relate to the mental and emotional stress and fatigue associated with learning to use, or using, the product. A product refined for good ergonomics will require very little effort to learn to use. It will also feel intuitive to use, thus causing very little emotional stress or fatigue.

See also

Development Reference: Design for Manufacturing (11.17), Design for Assembly (11.16).

Further Reading

North Carolina Department of Labor (2009) A guide to ergonomics. Available online at https://blueridge.edu/sites/default/files/pdf/continuing_ed/Aguidetoergonomics.pdf

Openshaw S, Taylor E (2006) Ergonomics and design: a reference guide. Allsteel. Available online at https://www.allsteeloffice.com/SynergyDocuments/ErgonomicsAndDesign ReferenceGuideWhitePaper.pdf.

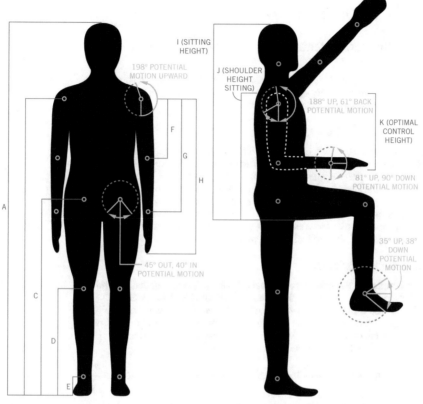

Dimensions and movement of the body. After (Tilley and Associates, 2002).

	WOMAN			MAN		
	99 Percentile	50 Percentile	1 Percentile	99 Percentile	50 Percentile	1 Percentile
A	69.8	64.0	58.1	75.6	71.3	62.6
B	55.4	50.4	45.4	60.6	54.6	49.1
C	36.6	33.4	29.6	40.1	36.5	32.5
D	19.7	18.0	15.8	21.7	19.8	17.3
E	3.3	2.9	2.5	3.7	3.2	2.6
F	11.0	10.4	9.2	12.3	11.0	9.7
G	20.7	18.6	17.5	23.1	21.1	18.9
H	28.5	26.5	23.5	31.5	28.6	25.5
I	37.1	33.8	30.8	39.3	36.0	32.7
J	25.2	22.0	20.8	26.6	23.6	20.8
K	14.8	13.2	12.5	16.2	14.4	12.5

*All Units In Inches

Body dimensions for the US population.

Ergonomics

11.26 Experimentation

Measure the performance of your design

Experimentation can provide high-quality information about a design. It can also be expensive in both time and money. To make the best use of these resources, it is desirable to do the following when experimenting:

Plan experiments before you do them

Carefully consider the experiments you intend to run. By thinking about all aspects of the experimental program, you can maximize the amount of information you will obtain. You should plan the number of test conditions, the number of repetitions at each condition, the kind of measurements to be taken, and the expected analysis of the data. When you do this before the experiment, you can make appropriate trade-offs to ensure you get the needed information within the time and resource limits. More information on experimental plans is found in Design of Experiments (11.18) in the Product Development Reference.

Test at multiple conditions

While it is tempting to test only at the optimal conditions, it is certain that the end user for a product will use it in less than ideal circumstances. In fact, you can probably count on somebody misusing the product in ways you haven't even imagined. Effective experimentation seeks to know how well the product performs under a wide variety of conditions. More information can be found in Robust Design (11.57) of the Product Development Reference.

Complete all non-destructive tests before potentially destructive tests

Prototypes are expensive, and sometimes not as robust as we would like them to be. It's tempting to avoid any tests that might harm the prototype. But as described above, it's important to test in non-ideal conditions. To allow completion of all possible tests, it's important to put off any tests that might damage or destroy a prototype until all other testing has been completed. Then, if the prototype is lost, all the rest of the information is still available.

For example, a team was developing a piece of carpet cleaning equipment that needed to survive a drop down a staircase. After completing the basic performance testing, the team held their collective breath and tossed the equipment down the stairs. Fortunately, it survived. In another case, a team developed variable output DC voltage supply. In their excitement to test the product after they got it working, they tested a high voltage to a low-impedance load. The resulting high current rush destroyed some of the components in the supply and required a time-consuming rebuild before testing could be completed.

Record all data and setup information necessary to perform the experiment

In the excitement to get testing, it can be easy to jump into testing without recording the test setup or the means of data collection. Sometimes only the output data is recorded. Environmental data (such as the temperature of the room) can easily be overlooked. In general, it is best practice to record anything you can think of that might affect the test.

It can also be helpful to photograph the test setup and the test in progress. Sometimes things that were not recorded can be determined by referring to appropriate photographs.

Experimentation

When mistakes are made in the setup or recording data, do not delete the mistake. Instead, indicate the error clearly, but keep the original record. This may help to avoid mistakes in the future.

Calculate preliminary results during experimentation

During the experimentation, while the data is being collected, a preliminary processing of the results (perhaps a calculation) should be performed. Preliminary processing can identify mistakes in experimental setup and prevent hours of wasted data. By processing as you go, you will know if you are achieving reasonable results.

It can be helpful to create a data calculation worksheet or template as part of the experimental plan. Then when the experiment is carried out, the output data can be entered into the template, the calculations carried out, and the results obtained almost immediately. These results are then evaluated to see if they make sense.

Analyze the data statistically

Every experiment has a degree of uncertainty or randomness associated with it. As experimental conditions change, the results of the experiment change. It is important to determine whether these changes reflect real differences in performance, or are simply due to chance or randomness. Statistical methods can be used to assess the significance of results.

In addition, statistical methods can be used to create approximations of the behavior of the product over a range of operating conditions.

The power of statistical analysis is greatest when the experiment has been planned from the beginning with statistical analysis in mind.

Prepare a written summary with at least one summary data plot

Careful thought about the test, results, and inferences should be captured in writing for transferability. If the results are only in the head of one team member, they cannot be used for any other purpose. In contrast, if they are thoughtfully written, they can be referred to again and again.

In many cases, the results of an entire experimental program can be summarized in one careful data plot. Such a plot makes it easy for one who is unfamiliar with the work to grasp the results. Often, it is worth thinking about the summary plot that is desired before the experimental plan is finished.

Following the above principles helps assure the most effective outcomes from experimentation.

See also

Development Reference: Dimensional Analysis (11.21), Robust Design (11.57).

11.27 Failure Modes and Effects Analysis

Take concrete steps to reduce the risk of product failure

Failure Modes and Effects Analysis (FMEA) is a subjective, yet structured, design tool aimed at identifying failure risk and reducing it through design changes. FMEA may be carried out on the product (termed design-FMEA or DFMEA), or it may be carried out on the process used to manufacture the product (termed process-FMEA or PFMEA). FMEA produces a Risk Priority Number (RPN), which guides the team to focus on the high-risk failure modes.

How to do it

1. Identify and list system components. Also list the functional purpose of the components; there may be multiple purposes for one component.
2. For each component, identify and list possible failure modes.
3. For each failure mode, identify the most severe potential effects of the failure.
4. Rate the severity of each effect, using the severity scale from the rating table.
5. For each failure mode, identify possible causes; there may be multiple causes for each failure.
6. Rate the likelihood of each potential cause, using the likelihood scale from the rating table.
7. Identify the controls preventing each cause, and the indicators that the cause exists before the failure effect occurs.
8. For each cause, given the controls and indicators from the previous step, rate the probability that the cause will be detected before the failure effect occurs, using the detectability scale from the rating table.
9. Calculate the Risk Priority Number (RPN) for each cause as the product of the severity rating, the likelihood rating, and the detectability rating.
10. For items with a high severity (9-10) or a high RPN (over about 125) take actions to reduce the RPN; after actions are complete recalculate the RPN to show improvement.

Tips

FMEA is generally carried out late in concept development or early in subsystem engineering. FMEA can be carried out on individual components, subsystems, or the system.

FMEA is best when careful justification is given for the subjective ratings. It is best to have a small team, rather than an individual, work on the FMEA. Where feasible, a multi-function team is often more effective than a single-function team.

Using a series of FMEAs can be an effective strategy for evolving the product. For example, a high-level FMEA can be performed to filter out the mundane many in order to focus a more detailed FMEA on the critical few.

See also

Lean Six Sigma's steps and templates for FMEA available at lssacademy.cpm/2007/06/28/10-steps-to-creating-a-fmea.

Development Reference: Fault Tree Analysis (11.28).

FMEA Rating Table

Rating	Description of Severity rating	Description of Likelihood rating	Description of Detectability rating
1	The effect is hardly noticeable to the customer	Extremely low; remote chance of occurrence	Almost certain to detect before failure
2-3	Slight effect that causes customers some annoyance, but they do not seek service	Medium low; slight chance of occurrence	High chance of detection before failure
4-6	Moderate to significant effect, customer dissatisfaction, service sought and/or required	Medium; occasional occurrence	Medium chance of detection before failure
7-8	Serious effect, system may not be operable; elicits customer complaint	Medium high; frequent occurrence	Low chance of detection before failure
9-10	Hazardous effect, complete system shutdown; safety risk; life threatening	High; regular occurrence	Very remote to no chance of detecting before failure

Grappling Hook

Climbing Rope

Gun To Launch Hook & Rope

In 2011-2012 the Air Force Research Laboratory sponsored a BYU Capstone team and asked them to design, build, and test a climbing system that would allow troops to scale vertical surfaces 90 feet in height. The device needed to be compact and carried by troops as standard equipment. The entire process of preparing to and climbing the surface should take only five minutes.

The team chose a rifle-mounted device that launched a hook attached to a rope to the top of the climbing surface. A battery-powered winch was then attached to the rope, and the winch lifted the solider to the top of the climbing surface.

Shortly after selecting this concept, the team performed an FMEA. The results of the FMEA are provided in the table. Notice that the current situation (columns 6-9) was evaluated early in the process (5 Nov 2011), assignments were made, and the evaluation of the improved situation was completed after the assignments were completed (15 Jan 2012).

FAILURE MODE AND EFFECTS ANALYSIS
For Product/Subsystem: Climbing Assistance Tool - Grappling Hook Concept, Rev. 2.0

Prepared By: Team Alpha
Date: 5 Nov.
Improved Situation: 15 Jan.

Component	Functional Purpose of Component	Failure Mode	Failure Effect	Failure Cause	Current Situation				Assigned Action	Improved Situation			
					S	L	D	RPN		S	L	D	RPN
Grappling Hook	Hold (Attach) Over Top Edge of Structure	Attachment Fails During Ascent	Climber Falls	Didn't attach well to top edge	10	4	6	240	Develop pre-climb pull procedure, Brady by Dec. 10	7	3	5	105
				Edge of Structure Fails	10	3	9	270	Develop pre-climb pull procedure, Brady by Dec. 10	7	3	5	105
	Holds 300 lbs.	Rope Detaches from Grappling Hook	Climber Falls	Poor Connector Design	10	4	8	320	Add redundant connector and perform FEA on new connector, Brady by Nov. 28	5	3	3	45
				Knot Fails	10	3	3	90	Add Redundant Knot, Bryan by Nov. 7	5	3	3	45
			Hook Unusable	Poor Material Choice	8	2	6	96	None	-	-	-	-
				Poor Geometry	8	2	4	64	None	-	-	-	-
				Hook Hits Object Head-on After Rifle Deployment	8	4	5	160	Add Calibrated Sighting Features to Rifle; Aaron/Jason by Jan. 14	8	3	3	72
Winch	Climb the Rope	Battery Exhaustion	Ascent Ceases	Poor Winch Choice	5	2	3	30	None	-	-	-	-
		Rope Slips in Winch	Climber Slips Down The Rope	Climber Using Wrong Rope	3	5	5	75	None	-	-	-	-
				Poor Pulley Design	3	3	3	27	None	-	-	-	-
		Winch Severs Rope	Climber Falls	Rope Jam	10	3	2	60	None	-	-	-	-
				Rope Wear	10	4	8	320	Procedure To Inspect As You Climb; Dave by Jan. 14	10	4	3	120
Rifle Deployment	Launch Grappling Hook 90 feet Vertically	Rifle Does Not Launch 90 feet	Will Not Be Able to Climb 90 feet	Hook and Rope Too Heavy	5	2	3	30	None	-	-	-	-
		Rifle Barrel Explodes Due To Attachment Weight	User Injury	Trying to Launch Too Much Weight	9	2	4	72	Trying to Launch Too Much Weight	9	2	3	54

11.28 Fault Tree Analysis

Understand potential causes of failure

Fault tree analysis (FTA) is a method of identifying possible failures in a product and coupling them with failures in the manufacturing process or in components or subsystems of the products. FTA helps provide probability estimates for individual failures, and provides failure information that supports FMEA.

Key methods

Fault tree analysis is a tree-based analysis that can be used to perform quantitative probability analysis on particular failures.

FTA starts with identifying a critical failure. All possible causes for the failure are then listed. The logical relationships between each of the causes are identified.

The next level of the FTA is then created by looking at all the causes for the top-level failure to identify the situations that would lead to these second-level conditions.

The process is repeated until it reaches the level of basic event causes, that is, those events that are the possible triggers for the high-level failure.

Quantitative analysis can then be performed on the fault tree. Given estimates of the likelihood of each of the basic events, the probabilities of each of the higher events can be calculated based on the structure of the fault tree. This can ultimately lead to an assessment of the probability of the top-level failure occurring.

Given a fault tree, one can see risk points for failure, which provide guidance for system redesign to increase reliability.

Applicability

FTA is generally performed during subsystem engineering. It is revisited during later development stages as problems occur.

A single fault tree thoroughly analyzes a single failure possibility. It does not analyze all possible failure modes for a product. Each critical failure mode must have a separate FTA completed.

See also

Development Reference: Failure Modes and Effects Analysis (11.27).

Further Reading

Wikipedia has an excellent article on Fault Tree Analysis: https://en.wikipedia.org/wiki/Fault_tree_analysis

NASA has prepared a fault tree analysis handbook: Vesely W et al. (2002). Fault tree handbook with aerospace applications. Available for download at http://www.hq.nasa.gov/office/codeq/doctree/fthb.pdf

Fault Tree Analysis

SR

PR

PRR

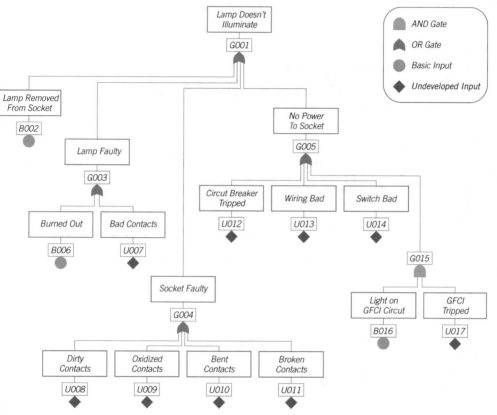

A fault tree for the lamp in a room failing to light. There are four direct causes for the failure to light: the lamp could have been removed from the socket, the lamp could be bad, the socket could be faulty, or there could be no power to the socket. Any one of these occurrences would cause failure to illuminate, so they are connected with an OR gate. One cause of no power to the socket would be if the lamp were on a GFCI circuit and the GFCI were tripped. Both of these events are necessary to create a lack of power to the socket, so they are connected with an AND gate.

Some inputs, like a burned out lamp, are basic inputs. Others, like the circuit breaker being tripped, have causes, but the causes are not considered in this analysis, so they are considered undeveloped inputs. A more complete fault tree analysis could develop the undeveloped inputs.

11.29 Financial Analysis

Understand financial implications of design decisions

Financial Analysis is a set of methods used to predict and track the financial implications of decisions made during product development. It is used to support decisions about what is to be developed. It is also used to support decisions about specific activities to be undertaken during product development.

Key methods

There are two commonly used systems of financial analysis for projects, payback analysis, and net present value. Payback analysis is simpler, and often used for short-term projects. Net present value is more complex and more correct, and is typically used for long-term, high-value projects.

Payback analysis

In payback analysis, the total costs of the project are calculated. The periodic (i.e., monthly) benefits of the project are also calculated. The total length of time until the sum of the benefits equals the sum of the costs is the payback period. It is common for companies to have a standard maximum payback period for funding project. For example, a company may say they will only fund projects with a payback of 18 months or two years.

When calculating benefits in product development, it is important to only calculate the net benefits. For example, if a product is sold for $10.00, with a manufacturing cost of $4.00, the benefit for selling 10,000 products per month is $60,000 per month, not $100,000 per month.

Net present value

Net present value is used to analyze projects with long lifetimes. Net present value calculations include the time value of money. This means that money in the future is worth less than money in the present. Therefore, future cash flows are discounted to reflect this difference in value.

In net present value analysis, the costs and benefits throughout the life of the project are calculated. Benefits (and costs) that occur in the future are converted to their present value through the discount equation:

$$PV(FV, t, r) = \frac{FV}{(1+r)^t} \tag{11.6}$$

where PV is the present value, FV is a future value that occurs t time periods in the future, and r is the discount rate, which is analogous to an interest rate. It is important the t and r are for equivalent periods, that is, if r is an annual interest rate, t must be in years. Or if r is a monthly interest rate, t must be in months.

We can also solve equation 11.6 for FV, r, and t:

$$FV(PV, t, r) = PV(1+r)^t \tag{11.7}$$

$$t(FV, PV, r) = \frac{\log(FV/PV)}{\log(1+r)} \tag{11.8}$$

$$r(FV, t, PV) = \left(\frac{FV}{PV}\right)^{1/t} - 1 \qquad (11.9)$$

When all the costs and benefits for the project have been converted to present values, the total present value costs are subtracted from the total present value benefits to give the net present value (NPV):

$$NPV = \sum_{i=1}^{N_B} PV(B_i, t_i, r) - \sum_{j=1}^{N_C} PV(C_j, t_j, r) \qquad (11.10)$$

where N_B is the number of benefit payments in the life of the project, N_C is the number of cost payments in the life of the project, B_i and t_i are the benefit amount and the time for benefit i, and C_j and t_j are the cost amount and time for cost j.

If the NPV is positive, the project is considered to be a good investment.

Projects should not be ranked by NPV, with the highest NPV considered the best project. The internal rate of return is the appropriate method for ranking projects according to their financial goodness.

NPV is sensitive to the discount rate. Choosing a discount rate should be done carefully. Most companies will have a specified discount rate. If there is no company-specified rate, the rate should be chosen to be approximately the return on investment for the company.

The use of discounted cash flows to analyze the financial performance of projects is a subject often called *Engineering Economics*.

Applicability

Financial analysis should be performed during the opportunity development stage of product development to determine broad parameters for success. The financial analysis should be revisited at each subsequent stage. Projects may be canceled if it makes no financial sense to continue with them.

See also

Ulrich, K.T. and S.D. Eppinger, *Product Design and Development*, 5th ed, 2012, McGraw Hill Irwin, pp. 354-372.

Development Reference: Cost Estimation (11.11), Cost Targets (11.12).

Financial Analysis

11.30 Finite Element Modeling

Obtain high-fidelity performance predictions

Finite element modeling (FEM) is an engineering modeling technique for obtaining numerical solutions to problems that are too complex for developing analytical solutions. FEM can be used for structural modeling (stress, deflection, vibration), thermal modeling (temperature, heat flux), fluid modeling (pressure, flow rate), electrical modeling (potential, current), and more. Models can also be coupled between the various domains.

Key methods

Finite element modeling consists of three fundamental steps: pre-processing, solving, and post-processing.

Pre-processing is the first step in finite element modeling. It consists of creating appropriate geometry, applying boundary conditions, and applying loads.

Creating geometry requires the creation of the overall geometry to be modeled, then crating a mesh that fills the model geometry. Different types of mesh elements can be used to give different fidelity models; higher-order elements will require more solution time, but generally more accurate results.

Solving involves executing a solver to solve the relevant partial differential equations. The geometry, boundary conditions, and loads are supplied to the solver and the solver calculates results. There is very little user interaction during the solution phase. Depending on the fidelity of the model, solution times can be seconds, minutes, hours, days, or weeks. Parallel computing can often be used to decrease the solution time.

Post-processing is the step that involves getting the desired numerical solutions out of the digital solution files and into a form that is usable for the design team. Post-processing might involve automatically parsing an output file to obtain a particular number or set of numbers. It also may include plotting contours or shaded surfaces to indicate stress or temperature levels. Even more complex post-processing is possible. Post-processing often requires the largest amount of user interaction time when performing finite element analysis.

Many CAD/CAM/CAE systems either include a finite element modeling package or support exchange of data with finite element packages. Sometimes the CAD system can be used for pre- and post-processing; other times it merely exports geometry data for use with a finite element pre-processing package.

In many cases, the geometry used for FEM needs to be simplified, because small features that have little or no effect on the behavior being modeled will cause very fine meshes to be created, with a corresponding increase in solution time.

Applicability

Finite element modeling is most often used during the subsystem engineering stage of development.

See also

Development Reference: CAD Modeling (11.6).

Further Reading

Zienkiewicz OC, Taylor RL, Zhu JZ (2005) The finite element method: its basis and fundamentals Sixth ed. Butterworth-Heinemann.

Bathe KJ (2006) Finite element procedures. Klaus-Jürgen Bathe, Cambridge, MA.

Meshfree methods are also available: http://www.compumag.org/jsite/images/stories/newsletter/ICS-07-14-2-Rodger.pdf

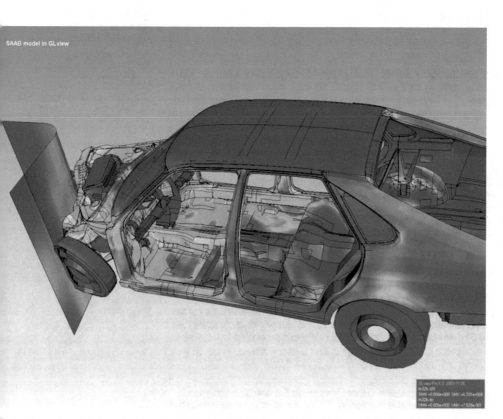

SAAB model in GLview

Finite element model of a vehicle crash. Image from Wikipedia, and placed in the public domain.

11.31 Focus Groups

Find out what the market wants in a hurry

A focus group is a group of 6-10 individuals who are asked a series of open-ended questions by a moderator in order to understand how they feel about a product or service. Properly done, focus groups can be a rich source of qualitative understanding about the desires of the market.

How to do it

- Determine the purpose of the focus group. The purpose of the focus group should be narrowly defined in order to guide the rest of the planning. Vague purposes will lead to vague focus groups that provide little useful information.

- Identify and invite the participants. The participants in the focus group should be selected to be appropriate for the purpose identified above. Homogeneous groups will tend to work together better and explore the subject more deeply. If heterogeneity is desired, it may be best to have a heterogeneous set of individually homogeneous focus groups.

- Develop a questioning route. A focus group will generally have from 8 to 10 questions. The questions should be carefully designed and tested before the focus group convenes. In general, questions should be open-ended, clear, short, and one-dimensional. The sequence of questions is ordered so that general questions come before specific questions, positive questions come before negative questions, and uncued questions come before cued questions. The time to be allocated to each question must be determined so all the questions can be covered in the focus group.

- Conduct the focus group. The focus group involves the participants and the moderator. The moderator is vital to successful outcomes in focus groups. The moderator must involve all of the participants, and must seek to draw out the participants' ideas, rather than reinforce the moderator's ideas. In many cases, it may be desirable to use a professional moderator to conduct focus groups.

  The focus group should be recorded (at least audio, preferably video) so that the participants' responses can be thoroughly analyzed.

- Analyze the data. The recording of the focus group should be transcribed. The transcription is then reviewed to identify common themes. This analysis will necessarily be somewhat subjective, so be very sensitive to biases among the analyzers. The analysis must seek to identify the participants' feelings, rather than seeking to confirm the design team's opinions.

Limitations of focus groups

- Focus groups are generally poor at determining quantitative data, such as numerical ratings of products. Surveys are much better for this type of data.

- Focus groups can tend to reflect groupthink, or be dominated by a single strong personality. Skilled moderators can help minimize this problem.

- Data from focus groups is difficult to analyze statistically, because it is qualitative rather than quantitative. This lack of statistical analysis causes some researchers to discount the use of focus groups.

- Focus groups provide little or no information about items that are not part of the question sequence. Be sure that you don't interpret the results more broadly than deserved.

Further Reading

One of the most cited books on focus groups is Krueger RA, Case MA (2009) Focus groups: a practical guide for applied research, 5th ed. Sage, Los Angeles.

A good checklist for conducting focus groups can be found online at https://assessment.trinity. duke.edu/documents/How_ to_Conduct_a_Focus_Group. pdf

See also

Development Reference: Interviews (11.34), Observational Studies (11.40).

Focus groups provide an interactive way of obtaining qualitative information from market representatives.

11.32 Goal Pyramid

Focus on the most important product development goals

In the early stages of product development the team identifies many requirements for a successful product. Because requirements differ in importance, it would be wrong for the team to give equal attention to all of them. The goal pyramid organizes the performance measures to help the team focus appropriately in order to create outstanding products.

- A few performance measures are *Constraints*. Constraints must be met in order to have a viable product, but there is not much concern about *how well* the constraints are met. Constraints can usually be recognized as yes/no performance measures.
- A handful of *Key Success Measures* distinguish between ordinary and great products. Typically there are about five to nine key success measures. If a product reaches the target or better for the key success measures, it will be considered a very desirable product. The development team's best efforts should be put toward achieving the targets for key success measures.
- There may be one or two *Stretch Goals* for key success measures. Stretch goals are values that exceed the target and would constitute exceptional performance. However, it is not clear that stretch goals can be achieved. Achieving stretch goals constitutes exceptional performance, but failure to achieve stretch goals does not constitute poor performance. The number of stretch goals should be limited to avoid diffusing the team's focus.
- Performance measures that are neither constraints nor key success measures are *Basic Performance Measures*. Basic performance measures must be met to have an acceptable product, but there is little value in optimizing to have performance be better than acceptable.

By clearly identifying key success measures and stretch goals, the team can keep the design focus where it makes the most difference. At any stage of development, the desirability of the design can be largely determined by evaluating the likelihood of achieving targets and stretch goals for the key success measures.

How to do it

The following steps can be used to build a goal pyramid:

1. Make a list of performance measures. The performance measures should be product characteristics that are important to the market.
2. Identify the key success measures. In consultation with market representatives, find out those things that would really delight the customer and make a big difference in customer satisfaction. Try to identify a handful of measures that the market really cares about.
3. Identify stretch goals. As you consider the possibilities for the key success measures, find a couple of things that would amaze the customer if they were reached, and aim to reach them.
4. Identify the constraints. In consultation with market representatives, find out which performance measures need only be met. Once met, the market is indifferent to (or has no concept of) the degree to which it is met. Resist the temptation to classify most of your performance measures as constraints; be sure the market really is indifferent to the level of performance for the measures.
5. Any performance measures that are neither constraints, key success measures, nor stretch goals are classified as basic performance measures.

Applicability

The goal pyramid is mostly created during Opportunity Development. In most cases, stretch goals are only identified during Concept Development, after an overall concept has been selected. The pyramid is revised as necessary throughout the product development process.

The goal pyramid and the requirements matrix

You can use the goal pyramid to organize the performance measures in a requirements matrix. Classify the performance measures into constraints, basic measures, key measures, and key measures with stretch goals.

All performance measures have acceptable limits. Unlike other performance measures, constraints have no ideal values. Therefore, the target value for a constraint will be a range that contains the acceptable limits. Only key measures with stretch goals will have a stretch goal as part of the real values.

Having classified the measures, it is effective to put key measures with stretch goals at the left of the performance measures list, followed by key measures, basic measures, and constraints. The real values in the matrix are also adjusted to match these categories. A new row of stretch goals can be added to the real values. A matrix with this structure facilitates focusing on the most important performance measures.

See also

Development Reference: Requirements Hierarchy (11.54), Requirements Matrix (11.55).

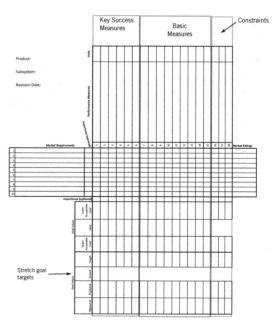

The goal pyramid graphically represents different kinds of performance goals for product development. The foundation consists of basic performance measures and constraints, which need to be in the acceptable region but don't drive purchase decisions. A relatively small number of key success measures distinguish between good and great products. One or two stretch goals can inspire the team to aim for an exceptional product.

The requirements matrix can be adapted to work with the goal pyramid by reordering the performance measures according to category and adding a row in the real values for stretch goals. Further, constraints have no ideal, and only a few key measures have stretch goals.

11.33 Internet Research

Access the world's knowledge from your desktop

Further Reading

Wikipedia's entry on data mining

Perhaps the most commonly used method of finding existing knowledge is a search on the internet. Because internet search engines can return millions of hits for a given topic, it is essential to separate the desired information from the noise.

It goes without saying that the internet is full of useful information. This is true even in the context of product development. For example, Amazon[1] can be used to access regional sales statistics for competitive products, product reviews, information regarding what other people also considered or purchased, and so on. This type of information can be extremely useful in establishing market requirements.

One of the greatest benefits of this type of information is that it is more current than what can be found in published literature.

Internet data can be accessed manually or by using data mining techniques. We recently used data mining techniques to search the internet for engineering design terminology. With relatively simple data mining techniques we searched 96,876 books in the Library of Congress, and 1,917,328 entries in Google Scholar. With the data mining techniques, the searches were completed in a matter of minutes.

For the Library of Congress data, shown in the figure on the opposite page, the data mining extracted the number of books published each year, where the search terms were present in the publication. It also extracted other interesting information such as the birth place of the author.

Internet searching can also be used to find existing solutions to a design problem. For example, a project was recently proposed to develop an improved portable dental chair by modifying an existing chair. A Google search of "portable dental chair" returned 806,000 results. On the first page of the results there were images of at least seven different commercially available chairs. There were three different sellers of multiple chairs. There was an article written by a non-profit world dental organization that described the strengths and weaknesses of various commercial chairs. In less than five minutes, the designer obtained a good understanding of the existing solutions.

One potential problem with internet searching is the assumption that all the good information is on the first few pages returned. In many cases, later pages have the best information. Deciding when it's time to end a search requires good judgment.

See also

Development Reference: Observational Studies (11.40).

Internet Research

[1] www.amazon.com

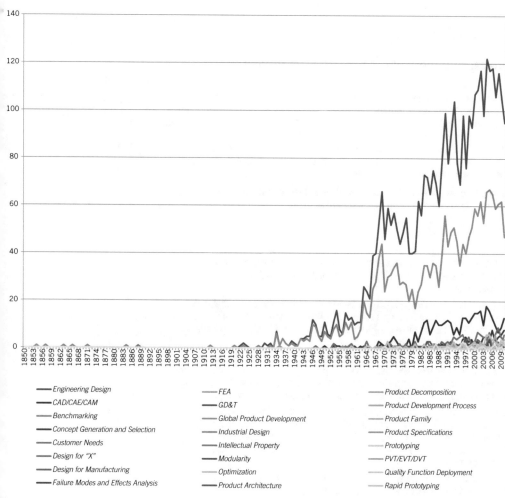

Legend:

- Engineering Design
- CAD/CAE/CAM
- Benchmarking
- Concept Generation and Selection
- Customer Needs
- Design for "X"
- Design for Manufacturing
- Failure Modes and Effects Analysis
- FEA
- GD&T
- Global Product Development
- Industrial Design
- Intellectual Property
- Modularity
- Optimization
- Product Architecture
- Product Decomposition
- Product Development Process
- Product Family
- Product Specifications
- Prototyping
- PVT/EVT/DVT
- Quality Function Deployment
- Rapid Prototyping

Internet research data collected with data mining techniques. Here the number of new entries added to the Library of Congress each year that include engineering design keywords.

11.34 Interviews

Get in-depth understanding of what the market wants

Interviews allow product development team members to have direct contact with people pertinent to the development project. Direct contact helps the team member get a firsthand account of needs, product use, opinions, complaints, difficulties, and so on. Interviews can be done with one, or with a few participants. One-on-one interviews are valuable because a participant's responses are uninfluenced by other participants. Small group interviews can be good because participants often have insightful dialogue with other participants.

A primary goal of interviewing should be to elicit an honest response from participants[1]. Don't be mistaken; this is very hard to do.

How to do it

- Establish what design question the interview will help answer: For example, how does a specific competitive product meet the market requirements of the current project?
- Choose who will be interviewed: Is there a certain demographic that needs to be targeted? Is there a sample population size the team will target? Griffin and Hauser (Griffin and Hauser, 1993) suggest that 90% of customer needs are identified after 30 interviews.
- Choose an interview location: This choice should be tied to what design question the interview will help answer. Generally, it should be as close to the context of the design problem as possible. For example, if we are designing camping stoves, we should be interviewing people in a campsite.
- Create and test the interview questions: The quality of the interview hinges on this point. Questions should be established with care. They should be tested and refined before performing the actual interviews. The questions asked should lead to the *type* of data the team needs (e.g., qualitative data, quantitative data).
- Determine how the interview data will be recorded: Options are notes (least disruption to the interview process), still photograph, audio recording, video recording (most disruption to the process).
- Perform the interviews: When performing the interviews, be flexible, look for hidden needs, and try to observe non-verbal information.
- Process the data and collect additional data as needed.

Starter interview questions

These questions may be useful in getting feedback about existing products on the market.

- When do you use this type of product?
- Why do you use this type of product?
- Walk us through a typical session using the product.
- What features do you like in this product?
- What features do you dislike in this product?
- What do you consider when buying a product like this?
- If there were no limitations, what would you do to improve this product?

Further Reading

Mulwa FW, (2003) Participatory monitoring and evaluation of community projects. Paulines Pub, Africa. See Ch. 5 – People-Friendly Evaluation Methods.

[1] Eliciting an honest response is the opposite of: (i) convincing the participant of what he or she needs, (ii) teaching the participant how to use a product that is difficult to understand, or (iii) making them think there is an answer you want to receive.

Interviews

OD

CD

See also

Development Reference: Observational Studies (11.40).

Interviewing a woman in Puno, Peru, about how she would use a small kiln to improve her ceramics business.

Interviews

11.35 Method 635

Generate more than 100 solution ideas in an hour or less

Method 635 is a group creativity technique that encourages more individual and collective thought to develop concepts at a deeper level than brainstorming. It was originally developed by Professor Bernd Rohrbach in 1968.

One of the benefits of Method 635 is that it engages quieter individuals on the product development team in a way that brainstorming typically does not.

How to do it

To carry out the activity, 6 people familiarize themselves with the problem at hand. Each creates and writes down 3 candidate solutions. The list of candidate solutions is then passed to the next participant, who adds 3 more candidates by building on the ideas from the previous participant. The built-upon-ideas are then passed to the next participant and eventually through all participants.

Hence the name 635, 6 participants, 3 ideas each, further developed by 5 other people.

By the end of the method (typically 30 to 60 minutes later), there will be 6 lists of candidate solutions, each of which contains 18 systematically developed candidates, for a total of 108 solution candidates.

If there are more or less than 6 people in the group, the process can still be followed, but the number of rounds and the number of ideas will be different.

See also

Development Reference: Brainstorming (11.5), Theory of Inventive Problem Solving (TRIZ) (11.66), SCAMPER (11.58).

Further Reading

Video: youtube, by linkmv97 called "Method 6-3-5 (Brainwriting)".

Method 635

CD

SSE

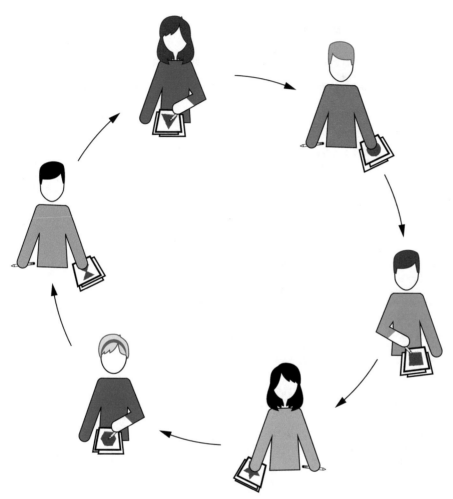

Six people each generate three ideas/concepts/solutions and pass to the remaining five people for refinement.

11.36 Mind Maps

Organize your thoughts to create new connections

Mind maps are visual representations of relationships between ideas. They are used to clarify thinking and to trigger new ideas. In product development, they are useful when creating concepts because they help increase the variety and novelty of the concepts developed.

Key methods

Buzan suggests that mind maps be done on paper, by hand. A number of software tools have been developed to create mind maps. This reference focuses on hand-drawn mind maps.

A mind map is created by starting with a blank piece of paper. Buzan suggests that ledger (11 inch x 17 inch) paper is preferable, because it allows plenty of space. It is recommended that color pencils or pens be available to make mind maps, because the colors will provide additional stimuli for the brain.

In the center of the blank piece of paper, a key word that summarizes the goal of the mind map is written, along with some kind of drawing that captures the essence of the key word. Words that are related to the key word are then written along lines that extend radially out from the center. The lines should be curved, rather than straight, according to the accepted wisdom of mind mapping. The lines should be different colors, and can taper, like tree branches.

For each of the related words, additional sub-words are created, fanning out like tree branches. Images can be added to key words. Words should be placed near related ideas. Sometimes there will be links between different branches. These can be indicated with dashed lines.

As the page fills up, you will capture a variety of thoughts related to the key idea. Some of the thoughts will have little direct value to your development project, but it is hoped that many will be useful.

The branches you develop on a mind map may prove helpful in developing concept classification trees and concept combination tables. They may also be helpful in decomposing the problem.

Applicability

Mind mapping is useful any time you want to create lots of ideas. It will often be used in the concept development stage.

Mind maps can be used when trying to find a way to solve seemingly intractable problems at any stage of development.

See also

Development Reference: Brainstorming (11.5), Method 635 (11.35), SCAMPER (11.58), Theory of Inventive Problem Solving (TRIZ) (11.66).

Further Reading

Tony Buzan, who is credited as the inventor of mind maps, has a website about mind mapping: http://www.tonybuzan.com/about/mind-mapping/

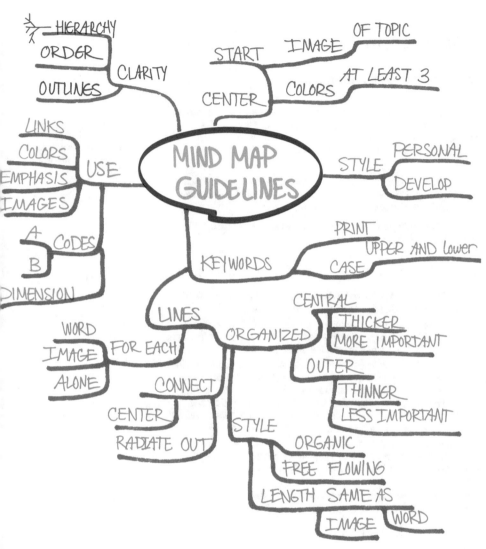

Mind map of guidelines for creating mind maps. After an image by Nicoguaro.

11.37 Multivoting

Quickly select a few good alternatives

Multivoting is a technique used by teams to perform a quick, intuitive selection of a few promising alternatives from a list of dozens of candidates. Although the evaluation is subjective, the use of many individual voters makes the process robust. Multivoting does an excellent job of capturing the collective feeling about the alternatives.

Multivoting can be used any time, there is a need to select a few items from a much larger list. It is particularly useful to select a manageable number of concepts for more detailed evaluation during concept development.

It is reasonable for a multivoting session to take less than an hour to reduce the concept set from 50 or more to 10 or less.

Further Reading

A brief handout of a formalized multivoting system has been published by the University of Kentucky at http://psd.ca.uky.edu/files/multivot.pdf

How to do it

1. Post a representation of all of the candidate ideas where they are visible to all team members. This is often done on a whiteboard or a wall.
2. Give each team member a fixed number of votes, generally between 5 and 15 votes per person. This can be done with stickers, adhesive notes, or just using the honor system and having members mark a vote with a pen.
3. Without talking, each team member votes for their preferred ideas by placing the stickers or marks on the representations, where all members can see them. Members should vote for concepts they want to carry forward. Members are not limited to one vote per concept; they may place all their votes on one concept if they so desire. The voting process is open, so members can watch the progress as they place their votes and adjust their voting as desired.
4. Count the votes for each of the candidate ideas. Those with few votes are eliminated. Those with high vote counts are carried forward. Those with low vote counts are dropped. If in doubt, it's probably best to carry a borderline idea forward.
5. If there are still too many concepts to move to the next decision stage, pursue another round of multivoting with fewer votes for each person.

In most cases, the goal of multivoting is to reduce the concept list to between six and fifteen concepts.

Applicability

Multivoting is done throughout product development as a means of quickly converging on a small set of ideas. At a minimum, multivoting is usually applied in the Concept Development stage.

See also

Development Reference: Scoring Matrix (11.59), Screening Matrix (11.60).

Section 5.4 of this book.

The results of multivoting for a set of concepts.

Multivoting

11.38 Nucor's Circles

Find market opportunities that match your skills and passions

Further Reading

Collins JC (2001) Good to great. Random House, New York.

Nucor's Circles can be used to systematically identify high-potential opportunities to pursue. Opportunities that lie at the intersection of (i) things we are passionate about, (ii) things we can be the best at, and (iii) things the market wants to have great potential.

How to do it

1. Choose one of the three sets: (i) things we are passionate about, (ii) things we can be the best at, and (iii) things the market wants. Make a somewhat large list of items for that set. For example, assume we choose the set *things we are passionate about*. The large list we create might include:

- High tech products
- Camping
- Phenomenal product design
- Helping people in need
- Materials
- Creating things in my shop
- Reading

2. Now choose another of the three sets and make a list while considering the already completed list. For example, make a list of *things we can be the best at* while considering *things we are passionate about*.

- Phenomenal product design
- Helping people in need
- Materials
- Hard work and dedication

3. Make a list for the remaining set. For this example, *things the market wants*. Do this in the context of the other lists.

- Helping people in need
- High tech products

4. Identify the items (if there are any) that exist in all three circles. These represent the opportunities of greatest potential.

Why it's named after nucor

Nucor Steel is considered to be one of the great Fortune 500 companies in an impressive study carried out by economist Jim Collins. Nucor significantly outperformed its competition in the stock market over a sustained period of time. When attempting to discover what distinguished Nucor from its competitors, Collins discovered that Nucor made all of its decisions in accordance with the overlapped section of three circles: passion, expertise, and economic denominators.

For Nucor the circles looked like: "Passion for eliminating class distinctions and creating an egalitarian meritocracy that aligns management, labor, and financial interests. Economic Denominator of profit per ton of finished steel. And could become the best in the world at harnessing culture and technology to produce low-cost steel."

See also

Development Reference: Project Objective Statement (11.49)
Chapter 4 of this book.

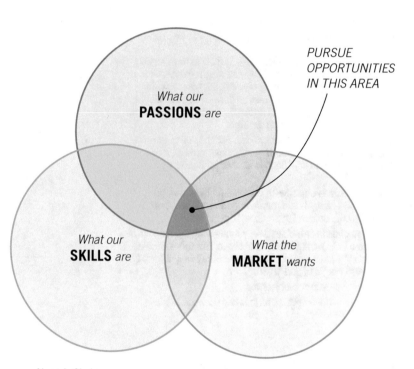

Nucor's Circles

11.39 Objective Tree

Determine the importance of multiple design objectives

The objective tree is a visual representation of the project or product objectives. It is a useful way to systematically establish weights that can be used in a scoring matrix.

How to create an objective tree

Consider the figure on the opposite page in the context of a horn design for a car.

1. Establish the top-level objective: This objective, labeled O1, is to have *a good horn design*. This is the top-level objective, so its overall weight (placed in the bottom right portion of the circle) in the objective tree is 1.00, obviously. Its weight relative to the objective one level up in the tree (termed relative weight, and placed in the bottom left part of the circle) is also 1.00 because there is no level higher than this top-level objective for this example.

2. Decompose the top-level objective into sub-objectives: For the horn example, we see three sub-objectives: Have good *sound quality*, have good *functional quality*, and have good *manufacturability*. Each is labeled for convenience (O11, O12, O13).

3. Assign relative weights to each sub-objective in the level (in this case the middle level of the tree): These weights are numbers between 0 and 1, and they sum to 1. They occupy the bottom left portion of the circles.

4. Calculate the overall weight of each sub-objective by multiplying the respective weight in this level, by the overall weight in the level above. For this example, the respective weight for O11 (0.30) is multiplied by the overall weight for O1 (1.00) to produce the overall weight of O11 (0.30).

5. Decompose mid-level objectives into sub-objectives.

6. Repeat previous two steps until decomposing further has no value to the project.

See also

Development Reference: Decomposition (11.14), Scoring Matrix (11.59).

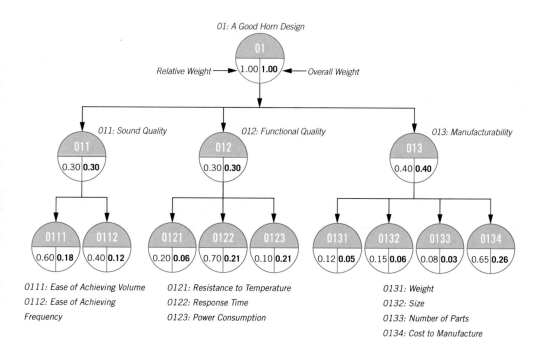

An objective tree for a horn design problem.

11.40 Observational Studies

Learn users' needs by watching their actions

Observational studies are one of the best ways to understand what people actually do in real contexts and time frames. What people actually do, and what they say they do are almost always different. For this reason, observational studies can be very valuable when trying to establish information that will guide the development process. An ethnography is a description of the customs of individual peoples and cultures; it is a natural artifact that emerges from observational studies. For example, having observed engineering students, we could say "Engineering students carry a scientific calculator in their backpacks. Most carry an extra set of batteries."

When designing products or services, we're interested in the meaning of activities and artifacts. We look for habits and rituals, and we keep an eye out for unanticipated issues and behavior.

Observational studies are most important because they help us ask "Why." Why do they carry the calculator in their backpack, and not on their wrist, or as a piece of software in their computer? Understanding why people do what they do helps designers to create objects that enhance lives without necessarily changing behavior.

Social science and anthropology are the disciplines typically trained in rigorous observational studies.

Further Reading

Lofland J, Snow D, Anderson L, Lofland LH (2006) Analyzing social settings: a guide to qualitative observation and analysis. Wadsworth. See especially Ch. 3 – Getting In and Ch. 5 – Logging Data.

Four observational study techniques

IDEO, a leading product development firm, produced a description of useful observational techniques (Ideo, 2003), a few of which are summarized here.

Rapid ethnography

Develop a relationship of trust with people relevant to the design topic. Do this by spending time together. Establish a relationship that allows you to visit and/or participate in the specific activities in the natural environment. This is termed a rapid ethnography because anthropologists often do non-rapid ethnographies that take years to complete.

Fly on the wall

Observe and record the behavior of people relevant to the design topic within their natural environment. Do this observation and recording without interfering with people's activities.

Behavioral archeology

Observe wear pattern, the organization and placement of things, and homemade solutions to problems. These are evidences of daily life, patterns of living, and patterns of thinking.

A day in the life

Observe users' behavior throughout an entire day. Record their activities and the context. Analyze the findings to discover patterns of behavior that only emerge over a sizable length of time.

See also

Development Reference: Interviews (11.34).

Observing users in their native environment is a great way of understanding market needs, including unstated needs.

One Pager

11.41 One Pager

Convey design decisions or key findings in a brief format

Product development teams create a substantial amount of information during the development process. It is important to recognize that although valuable, not all of that information play a *key* role in the development process. Key information should be preserved in a carefully prepared document and shared with the development team and others.

An extremely effective practice is to condense the key information into a single-sheet document that conveys the information in a clear and accurate way. We call these documents *one pagers*.

Principles motivating the use of one pagers

- A condensed description of your work and its outcome is more likely to get noticed and shared than the full description of the work. Simply stated: supervisors want you to share your message concisely.

- Once a condensed version of your work is noticed and appreciated, others will be more likely to engage the full description of the work.

- When the condensed version of your work has no substance or detail to back it up, it has no value.

- Selective use of one pagers raises their stature, which ultimately allows you to share your most important messages effectively. Overuse of one pagers relegates them to more unread paperwork.

Characteristics of good one pagers

- One pagers are short. If a "one pager" needs a staple, it's not a one pager.

- One pagers are navigable. Headings make it immediately obvious to the reader what the core sections in the one pager are.

- One pagers are meaningful. They contain information others want to know. This is different than simply containing information you want to share.

- One pagers are correct. If one pagers are to be trusted, they need to be absolutely 100% correct. When they are consistently correct, one pagers become highly valued.

- One pagers are objective. They represent the facts and your clear recommendation regarding them. They tell it as it is, without mincing words.

- One pagers are supported. Every element of the one pager could be backed up by a more comprehensive report of your work.

Memo versus one pager

In a sense, one pagers are like memos; they convey information to others. But the expectations for a one pager are much higher. Memos are often overused, one pagers are not. Memos often lack the good characteristics defined above. In fact, one pagers that do not have the characteristics listed above are simply memos, and unfortunately get added to the list of undervalued often unread paperwork.

OD

CD

SSE

SR

PR

PRR

ATL
TECHNOLOGY
"Precision Electronic Assemblies"

Desktop Cable Assembly Latch Failures and Proposed Solution

C. Mattson, M. Darrington, and A. Robison
02 July 2004

In this short report, we describe (i) the observed Desktop Latch failures, (ii) the root cause of the failure, and (iii) a proposed solution to eliminate or reduce these failures.

Succinct Text

The Observed Failures

As shown in Figure 1, the desktop latch can be crushed when the desktop cable assembly is mated with the viewer connector. Figure 1 shows the desktop cable assembly on the left, and the designed and crushed latches in the middle and right, respectively.

This failure is more frequently observed when the cable assembly is inserted into the PDA as shown in Figure 2. This insertion approach is one where the guide posts on the right side of desktop connector are inserted first, followed by the post and latch on the left.

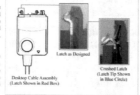

Latch as Designed

Crushed Latch
(Latch Tip Shown
in Blue Circle)

Desktop Cable Assembly
(Latch Shown in Red Box)

Figure 1: The Observed Failure (Right Picture)

The Cause of the Failure

The latch is crushing because the latch tip (see blue circle on right side of Figure 1) is exposed to forces sufficient to cause latch buckling and yielding. A cross-section of the desktop cable assembly (shown in Figure 3) confirms that the latch tip can be subjected to an undesirable force vector. The latch tip catches on the viewer connector; when it does so it cannot move inward (to the left in Figure 3), as it needs to do to prevent failure.

Proposed Solution

To prevent this failure, we must keep the latch tip from being exposed to undesirable forces. We generated and examined 14 different concepts that attempt to do this. The concept of adding a chamfer to the viewer connector, in the area where the latch is catching on it, is the one that we believe to be the most tool safe and schedule safe concept. We made a representative prototype, and found that it did not fail as described above.

Figure 2: Insertion Angle that Increases
Frequency of Failure

ATL Technology, L.C.

Meaningful Content

"Precision Electronic Assemblies"

> guide post holes (not the power plug hole). A
typed this connector and have cycled it 5000 times.
nector contact (strike plate for the latch) has not
art to fail, as the current design fails. More thorough
he modified tool.

Added Chamfer

Current Design: Catches   New Design: Slides

Figure 4: Current Design (left), Proposed Design (right)

2.06 mm

0.45 mm

0.57 mm

Figure 5: CAD Model of Suggested Change

Pertinent Navigation

Latch tip catch-
es on viewer
connector,
resulting in a
force vector that
can cause the
latch to buckle
and yield.

Force

Figure 3: Cross-Section Confirming that Latch Tip
may be Exposed to Undesirable Forces

Clear Technical Images

ATL Technology, L.C.

One Pager

An example of a one pager used to quickly convey proposed design changes. Once the basic message was understood and appreciated, the proposal was approved and a formal change to the design documents (engineering drawings, bill of materials, etc.) and the tooling was initiated.

11.42 Optimization

Get the most performance out of your product

Numerical optimization is a popular tool for computationally finding the highest performing designs. Importantly, optimization techniques not only tell you the best performance that can be achieved, but also the design that achieves it.

Optimization requires a mathematical model of the design's performance as a function of the design variables and parameters. Generally these models are not available until after the concept development stage. Therefore numerical optimization is typically used during subsystem engineering. By the time models are good enough to be used in an optimization setting, roughly 5-10% performance improvement can be gained.

The key to successful optimization: the problem formulation

Formulating the problem is by far the most important part of numerical optimization. The principle behind this is that the computer will search for exactly what you tell it to search for – even if it looks for unrealistic or physically impossible design features such as beams with negative lengths or holes that are bigger than the part it is supposed to be drilled into. A well-developed formulation makes the search meaningful.

Once a correct problem formulation has been made, virtually any solver, using any algorithm will result in performance improvements. A problem formulation looks like this:

$$\min_{x} \quad \mu(x,p)$$
$$\text{subject to} \quad g(x,p) \leq 0$$
$$h(x,p) = 0$$
$$x_{lower} \leq x \leq x_{upper}$$

where x represents the design variables, and $\mu(x,p)$ represents the design objectives, which are functions of variables (x, things the optimizer will change) and fixed parameters (p, things the designer chooses but does not let the optimizer change). The computational search for the variables that optimize the objectives will most often be subject to constraints on the search. These are represented by inequality constraints ($g(x,p)$), equality constraints ($h(x,p)$), and side constraints on the variables (x_{lower} and x_{upper}), which are simply the limits defining the acceptable variable values. It is useful to note that x, μ, g, h, x_{lower}, and x_{upper} can be scalars (when representing only one thing) or vectors (when representing multiple things).

As with any mathematical representation, it is valuable to understand what it represents in plain language. The generic formulation above says: find the values of the variables x that minimize the objective function μ, subject to inequality constraints, equality constraints, and side constraints on the variables.

How to create and solve the problem formulation

1. Choose design objectives. For example, minimize beam deflection or minimize stress. It is common to express them all as minimization statements, though this is not necessary. To maximize the objective function using the form above,

Further Reading

Messac A, (2015) Optimization in practice with Matlab. Cambridge Press.

For multiobjective optimization: Balling, RW (1999) Design by shopping: A new paradigm?. WCMSO-3.

Messac A, Ismail-Yahay A, Mattson CA (2003) The normalized normal constraint method for generating the Pareto frontier. Struct Multidiscip O 25(2).

simply minimize the negative value of the objective function (i.e., min $-\mu$), or to reach a target simply minimize the squared difference between the optimal design and the target design (i.e., min$(\mu_{target} - \mu)^2$).

2. Choose design variables (x) that the optimizer will change. For example, the lengths, widths, and heights of a beam. Also choose the fixed parameters that the optimizer won't change. For example, Young's Modulus.

3. Identify and set the constraints of the problem. For example, the beam bending stress ($\sigma_{bending}$) must be less than the yield strength (S_y). Many optimizers expect to see constraints in a certain form such as the one shown above ($g \leq 0$). For the beam stress constraint simply move the terms around; $\sigma_{bending} \leq S_y$ can be expressed as $\sigma_{bending} - S_y \leq 0$. Also establish the upper and lower acceptable limits for the variables (x).

4. Choose and use one of many search algorithms. Choices include (i) *gradient-based optimization* algorithms, (ii) heuristic methods such as *genetic algorithms*, or (iii) *brute force optimization* methods. A simple internet search for these terms will lead to valuable information.

5. Evaluate the optimization results to see if a meaningful result was achieved. Adjust the problem formulation as needed to be more meaningful and repeat these steps.

Single objective versus multiobjective optimization

When there is only one objective (μ_1) in the problem formulation, a single optimal design results. On the other hand when there are multiple competing objectives (μ_1, μ_2, ..., μ_n,), a set of optimal designs results. The set defines the optimal trade-off curve between the competing objectives. This curve is called the Pareto frontier and it contains designs that can only be improved in one objective by giving up in another objective.

See also

Development Reference: Robust Design (11.57), Sensitivity Analysis (11.61), Uncertainty Analysis (11.68).

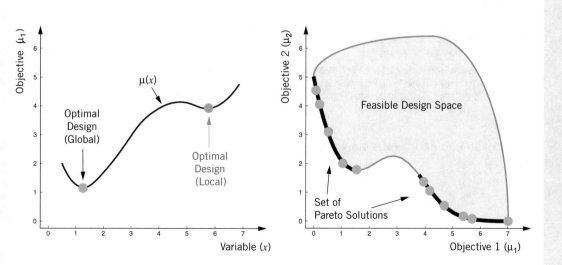

Optimal solutions for single objective problems (left side) and multiobjective problems (right side).

11.43 Patent Searches

Find creative solutions others have found for similar problems

Patents provide an excellent source of creative ideas for solving challenging problems. They also provide information about designing products to avoid infringement.

Patent searches are used as one way to find out what has been done relating to the product being developed. During early stages of development, patents are searched to find benchmark concepts and inspire creativity. During later stages of development, patents may be searched to avoid infringement of existing patents and to determine whether a new idea is patentable.

Key methods

Google Patents is an excellent online tool for searching patents. It allows searching by patent number, by classification, by issue date, by inventor, and by keywords in the patent title, abstract, and body. The results of the search are available for download as .pdf files.

A general strategy for doing a patent search is to use keywords to find relevant patents. Once relevant patents have been identified, there are three ways to expand your search. First, you can find all the patents that are cited as prior art in the patent you found. Second, you can find all the patents that cite your patent as prior art. Third, you can search the classifications of your patent.

As you search using these methods, the number of patents obtained will quickly expand. You will need to develop a skill at quickly examining the abstract, images, and claims to see which patents are relevant to your development project.

You should recognize that expired patents can provide sources of ideas as readily as current patents. Furthermore, expired patents can form the basis of prior art that can help you avoid infringing current patents.

Applicability

Patent searching is useful in the concept development stage, both as a way to see how others have solved your same problem and as a way to expand your understanding of related problems.

The presence of a valid patent in your development area does not mean your project is doomed. You may be able to develop means of avoiding patent infringement. Alternatively, you may be able to license the patent with which you are concerned.

See also

Development Reference: Internet Research (11.33).

Further Reading

Google has a very flexible and powerful patent search capability: http://www.google.com/?tbm=pts

Five different uses for patent searches in product development are identified here: http://ip-updates.blogspot.com/2004/07/patent-searching-in-product.html

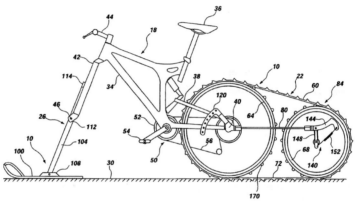

Image from US Patent 6663117 B2, which resulted from the work of a Capstone team at BYU.

11.44 Performance Measures

Measure how well your product meets the user'S needs

The market requirements represent the things the development team has found out that the market wants in the product. As the team is *not* the market, the team cannot make a direct judgment about the desirability of the product. The team must develop measurable characteristics of the product that can be predicted and measured by the team throughout the product development process. These measurable characteristics are *Performance Measures*.

Characteristics of performance measures

Effective performance measures share the following characteristics:

1. Concept independent:

2. Unambiguous (even if subjective):

3. Related to market requirements:

4. Dependent, rather than independent:

5. Appropriate units:

6. Minimize use of yes/no:

How to do it

The following activities can be helpful as you develop an appropriate set of performance measures.

1. Review customer statements.

2. Ask yourself how you could measure each of the market requirements.

3. Imagine how you would measure the performance of two different products to determine which was best.

4. Search for often-used decision criteria for similar products on the market.

Applicability

Performance measures are developed primarily during opportunity development. Ideal values for performance measures for the system are identified during opportunity development. Ideal values for subsystems, and target values for the system and subsystems are identified during concept development. Measured and predicted values are obtained during subsystem engineering and the product refinement stages.

Performance measures may evolve throughout the design process. Changes in the performance measures should be tracked with the ECO process.

See also

Development Reference: Requirements Hierarchy (11.54), Requirements Matrix (11.55).

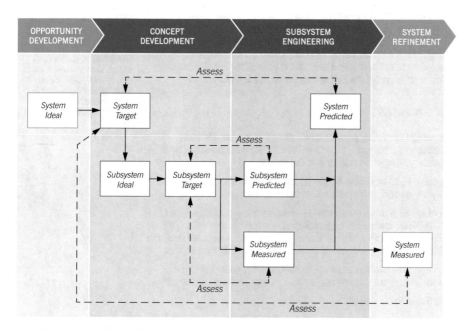

Real values for the performance measures evolve during product development. Target values for systems and subsystems are identified during concept development. Predicted values for the system, and measured and predicted values for subsystems, are obtained during subsystem engineering, and are assessed by comparing with the appropriate target values. During system refinement, measured values for the system are obtained and compared with the targets. As the design continues to evolve, measured and/or predicted values will likely change as well. Each time they change, they should be compared with the target.

11.45 Personas and Locales

Understand and teach others about who your product serves

Personas are archetypes of the people – often end users – who will be affected by the product you're developing. They're thoughtfully developed by the product development team to better understand who the product serves, and to teach other decision makers about those people. Effective use of personas helps the team abandon the idea that their product will serve 100% of the population, helps them focus on who the product does serve, and when used during the decision-making process, helps teams create products that are truly loved by the market.

How to create a persona

1. Establish the objective of your product development effort (see Project Objective Statement). This is an important first step because personas need to be developed for your specific project and its objectives.

2. Considering the project objective, decide which general part of the population is served by your product (e.g., people with difficulty sleeping).

3. Discover as much as you can about the general part of the population your product serves. While doing this, begin noticing and further researching the characteristics that are pertinent to your product. Various sources are available such as online databases, social media analytics, web analytics, and more.

4. Look for trends in the data that allow you to start dividing your part of the population into primary and secondary users and other pertinent stakeholders.

5. Develop 3 to 5 *believable* personas:

   - Give each persona a name

   - Choose a photograph to represent the persona

   - State the role of the persona (e.g., primary user, maintenance worker)

   - Develop a description of the persona including their desires, needs, expectations, motivations

6. Evaluate the persona and revise as needed.

How to use personas to advance the design

Develop and use believable personas early in the product development process to help team members be human-centered during the development. Post the persona description on the wall of the team space. Refer to it often. Use a persona during a team meeting and see how it changes the discussion from that of *product* to that of *who the product serves*. This can be particularly useful when team decision-making digresses to a struggle between what one team member wants versus another. Personas help teams shift back to what the end user would want.

Persona pitfalls to avoid

To be useful, personas need to be accurately developed. They cannot be based on team member opinions about who the end user *might* be or based on unvalidated assumptions. Personas need to be believable before they can become something the team rallies around.

Further Reading

van Boeijen A, Dallhuizen J, Zijlstra J, van der Schoor R (2014) Delft Design Guide BIS Publishers. pp. 94-95.

Hanington B, Martin B. (2012) Universal methods of design: 100 ways to research complex problems, develop innovative ideas, and design effective solutions. Rockport Publishers. pp. 132-133.

Locales

When the location where the product will be used is crucial to the product design, a locale can be developed. The locale is a description similar to a persona, but it focuses on a real or representative location. It has the same product development objectives as a persona, but with a focus on location and context for product use. The use and pitfalls of locales are the same as personas.

See also

Development Reference: Project Objective Statement (11.49), Storyboards (11.64).

Persona representing a shopkeeper in Puno, Peru.

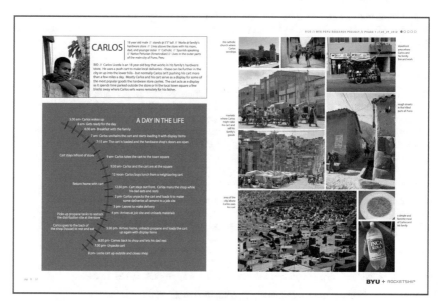

Locale describing Puno, Peru.

Personas and Locales

11.46 Plan Do Check Act (Shewhart Cycle)

Formalize the improvement of product quality

Plan Do Check Act (PDCA) describes the four parts of the Shewhart Cycle for statistical quality improvement. PDCA is used to improve the performance of repetitive processes, such as product manufacturing. It can also be used to improve the design of products.

The four steps in the Shewhart cycle are Plan, Do, Check (or Study), and Act. They describe a series of steps necessary to verify and formalize improvements in processes.

How to do it

First, identify a focus for the cycle, which is generally aimed at improving a product or process. The plan step of the cycle requires developing a hypothesis about how the process could be improved and planning an experiment to prove or disprove the hypothesis. The hypothesis should be written clearly, and should include measurable terms that can be evaluated during the check step. The plan for the experiment can be simple or complex. The experiment can be one that will take a few minutes or several weeks. The important thing is to plan the experiment before performing it, so that you can be sure the experiment will address your hypothesis.

The second step, do, requires carrying out the experiment that was planned in the first step. It seems so obvious that some people wonder why it needs a separate step all its own. But many times a company or individual will never finish an experiment they start because of external pressure, or a perception that other things are more urgent, or because initial results seem so promising that there is no need for further testing, or even because it's just "common sense." However, only through carrying out the full experiment as planned can the hypothesis be appropriately resolved.

The third step, check (or study), requires analyzing the results of the do step. Statistical methods should generally be used, in order to assess whether the changes (if any) observed are likely to be real, or just the result of randomness. Furthermore, it is desirable to study the results of the experiment deeply, as they can suggest ideas for future experiments in later PDCA cycles.

The fourth step, act, requires taking action on the system being studied that is consistent with the results of the study. If the study shows that the suggested changes make a significant improvement, the system must be adjusted so that the changes will be consistently implemented in the future. If the study shows that the suggested changes have no effect or make things worse, the results are documented so they can be transferred to others.

The cycle is then repeated by moving to the plan step for another potential improvement. In this way, the system is continuously improved.

Applicability

PDCA is most often used during the sustaining stage of product development.

See also

Development Reference: Design of Experiments (11.18).

Further Reading

Langley GJ, Moen R, Nolan KM, Nolan TW, Norman CL, Provost LP (2009) The improvement guide: a practical approach to enhancing organizational performance, 2nd ed. Jossey-Bass, San Francisco.

A classic guide to improvement using Plan-Do-Study-Act.

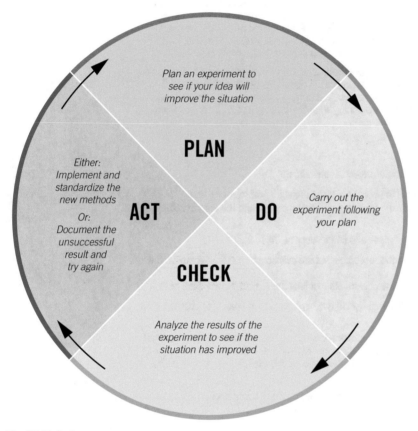

The PDCA Cycle.

11.47 Planning Canvas

Develop plans that cover all the bases

Planning is a fundamental part of every stage of product development. It's carried out at various degrees of formality, from detailed complex planning that involves many people, to the subconscious planning that occurs naturally in a designer's mind. No matter the case, five basic pieces of information are needed to create a good plan. These are shown as bold headings in the example canvas on the next page. The canvas is designed to help you cover your bases while planning.

A good test is to pause from your design work and fill in each of the areas of canvas. If you can confidently fill in the areas, you're right to be doing design work. If you cannot confidently fill in the areas, you've started your design work prematurely. Your time will be better spent planning so as to make sure your design work will produce a valuable outcome.

How to do it

First, scope the project, sub-project, or task. To do this:

1. Discover and state the objective of the project, sub-project, or task.
2. Discover and state the starting point for the project and the ending point you hope to get to.

Second, find out what enables and limits success. To do this:

3. Discover and state the resources that are available to you to accomplish the objective.
4. Discover and state the constraints that limit the project, sub-project, or task.

Third, choose actions that make good use of resources and respect constraints to meet the objective.

5. Establish and list one or more actions that would be needed to transition from the starting point to the ending point. When a team member is assigned to do or lead the action and important dates are added to the list, then a plan has been created.

Fourth, evaluate the completeness and accuracy of the information in the canvas, and the quality of the plan. Refine and update as needed.

Why it's called a canvas (and not a template)

Canvases are popular in business management. They are tools to visualize and assess a changing plan (such as a business plan or a business model). They are much more than a template. They are a working board. The name Canvas implies change and iteration, which is exactly what you'll need when it comes to planning. You'll need an initial plan based on the best information you have at the beginning. Then as more information is gained, you'll need to update that plan. Doing those updates directly on a canvas, or at least in the language and form of a canvas, will help the team see the plan is becoming more and more refined, as opposed to being frustrated that there always appears to be a new plan.

See also

Development Reference: Critical Path Analysis (11.13), Plan Do Check Act (Shewhart Cycle) (11.46), Value Engineering (11.69).

Revision: Jan 4, R2

1 Objective

Prepare for and drill a 4 foot borehole in cobblestone. Simulate real conditions.

2 Starting Point

Jan 4, we have a working system that was used to create a two foot hole in clay soil with no rocks. The slurry pump is inadequate, and we have no way to extract drill pipe yet.

3 Resources

- $500 for prototyping
- 110 man hours for design and build
- 32 man hours for the drilling day
- Relationship with landowner that may let us dig on his property
- Project sponsor availability (to bring realistic evaluation to the drilling day)
- Tools, computing, etc

5 Key Actions, Dates, and Responsibilties

- By Jan 10, Schedule the dig time with team and sponsor (Nathan)
- By Jan 15, Secure a place to drill (Elena)
- By Jan 20, Get gas company clearance (Elena)
- By Jan 20, Design a structure to extract pipe (whole team, led by Colleen)
- By Jan 26, Build a structure to extract pipe (Colleen, Nathan, and Eric)
- By Jan 26, Build an additional pump that can be used in series with existing pump (Jason and Sabin)

- On Day of Dig, Gather the needed equipment (Elena)
- On Day of Dig, Perform the dig (Whole team, minus Sabin)
- On Day of Dig, Collect data during the dig (Nathan and Chris)

- 3 Days after Dig, Compile & analyze the data (Nathan)
- 3 Days after Dig, have a team de-brief on what was learned at the dig (Whole team, led by Elena)

4 Constraints

- Must get approval from gas company
- Must dig on a Saturday, because of time conflict
- Need to use human power only

Ending Point

By Jan 30, we want to have a working system that sucessfully cuts through four feet of soil with cobble stones. Where we simulated real consditions for slurry pumping and pipe extraction

Planning Canvas

Planning Canvas for a small portion of the Human-Powered Well Drilling Machine project.

11.48 Product-Focused Requirement Statements

Obtain high-quality statements of product requirements

Product-focused requirement statements are requirement statements derived from user statements about the product that have been rewritten to be of most use in the product development process.

How to do it

To create a product-focused requirement statement, we begin with a statement from a user about what would be desirable in a product. The user statement is then rewritten so its expression has the following attributes:

Positive, not negative, phrasing: Sometimes the user will say what they want the product *not* to be or do. Rather than capturing what would be undesirable in the product, it is more effective to express positive characteristics of a desirable product.

As specific as the user statement: It is important to capture the user's intent as faithfully as possible. Teams will want to generalize user statements, but the generalization should happen later, after related requirement statements have been grouped.

A requirement of the product, not the environment or the user: Some user statements will be expressed in the form of statements about the environment, rather than about the product. Since the team is focusing on designing the product, requirements should be expressed for the product.

A requirement, rather than a performance measure: Users will often make statements that define performance measures and desired values of the measure. These performance measures should certainly be captured in the requirements matrix. However, the user-stated desired value is only one data point, and should generally be used to help define the ideal values, rather than being used as the ideal value. Furthermore, the team will learn much more about the requirements if the user is asked *why* the performance measure is important to the user. Often one or more requirements will be discovered by probing the reasons for stated performance measures.

A requirement, rather than a product feature: Users who have thought carefully about products will often have specific suggestions about ways to achieve desired performance. When user statements suggest specific features, it is more helpful to the design team to identify the requirements behind the suggested feature. By doing this, the team is free to consider other ways to meet the requirement, which may be more effective than the user's suggestion.

Independent of importance: As users express opinions about the product, they often include statements about the importance of various requirements, features, and performance measures. These statements of importance are valuable and should be captured by the team. The requirements matrix has entries for the importance of both requirements and performance measures. User statements of importance should be considered as the team determines relative importance of the requirements, rather than be directly included in the product-focused requirement statements.

See also

Development Reference: Requirements Hierarchy (11.54).

Sample requirement statements created from customer statements about a cordless circular saw. These statements illustrate the six guidelines of effective product-focused requirement statements.

Guideline	Customer Statement	Good Requirement Statement	Inferior Requirement Statement
Express the requirement with positive, not negative, phrasing	It just doesn't have the torque of a corded saw and will choke in a slight bind while cutting plywood sheets.	The saw resists binding when cutting plywood.	The saw doesn't bind when cutting plywood.
Express the requirement as specifically as the user statement	I wish it had a case to protect it from dents and dings	The saw is resistant to dents and dings	The saw is rugged
Express the requirement as a requirement of the product, not the environment or the user	The trigger safety interlock is aggravating! I have incredibly large hands and still find the trigger interlock a stretch to reach.	The saw can be started easily by most people, even with safety interlock features active.	People can easily reach the safety interlock.
Express a requirement, rather than a performance measure	The battery lasts for three hours of typical carpentry work.	The battery lasts a long time. *Or* the battery life allows typical carpentry work.	The battery lasts for three hours.
Express a requirement, rather than a product feature	Having the blade on the left side gives a right-handed user a great line of sight.	The saw provides clear sight to the cut.	The saw has the blade on the left side.
Express the requirement independent from its importance	The saw does not have a light which I miss.	The saw illuminates the cut area.	The saw should illuminate the cut area. *Or* the saw must illuminate the cut area.

11.49 Project Objective Statement

Explain your project objective in 25 words or less

A project objective statement is a clear and concise articulation of the project and its goals. It is an overarching mission statement and while simple, it is a powerful way of uniting a team. When the details of the project become overwhelming, the project objective statement serves as a firm mission the team has agreed to.

Characteristics of a good project objective statement

- Clear: The statement should be unambiguous about *what* will be done, *when* it will be completed, and *how much* it will cost (in resources) to do it. It should be written in a language that can be understood by all people interested in the outcome of the project.
- Concise: The statement should be short, approximately 25 words. As such, it should describe only the top-level scope, schedule, and resources. The conciseness of the statement is not easy to achieve, but this requirement forces the team to articulate the essence of the project with carefully chosen words.
- Informed: The statement is constructed based on information learned from the client or other stakeholders. It is not formed by the team in isolation.
- Validated: The finished statement is validated by the client or other stakeholders.
- Unifying: The finished, validated, project objective statement unites the team.

How to make and use a project objective statement

Prepare a project objective statement early. Do this as one of the first actions taken at the beginning of a project. It will help the team understand what is required, and it will help the client clarify in his/her own mind what is wanted. One team member, typically a team leader, drives the effort to create the statement. The driver seeks feedback as appropriate from the client and other team members until a final project objective statement is converged upon.

The project objective statement has little lasting impact if it is not frequently referred to and updated as necessary.

Example project objective statements

For a human-powered water well drill:

> Design, build, and test a human-powered drill that reaches underground potable water at depths of 250 ft in all soil types by March 25, 2011 with a prototyping budget of $2,800 USD and for less than 1,700 man hours of development.

For a manufacturing process that makes medical devices:

> Design and demonstrate a repeatable manufacturing process for a continuous-wire cannula to reduce variability and increase overall output within existing product specifications by March 31st, 2010 for under $1,200.

For a machine used to direct the RF beam of a military communication systems:

> Design and build a functional prototype control system and housing to actuate dielectric wedges and steer an RF beam with a budget of $3,500 by April 8, 2009.

See also

Chapter 4 of this book.

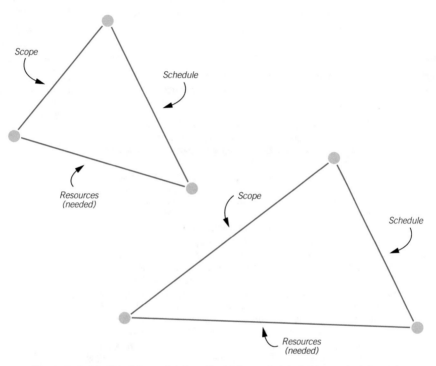

The project objective statement defines the triple constraint of scope, schedule, and resources for the project. Assuming the vertices must stay connected, this image shows that when one item (scope, schedule, or resources) expands or contracts, at least one other must also expand or contract.

11.50 Prototyping

Advance your design by building and testing prototypes

Prototyping is the act of physically constructing an approximation of a product or part of a product. It is a powerful way to learn and communicate during the product development process. As expressed by well-known design engineer David Kelley "If a picture is worth a thousand words, a prototype is worth a million." Prototypes are valuable because they capture physical phenomena that we often don't yet know how to model, or don't have time to model.

Types of prototypes

Physical prototypes exist in many many forms. Some can be made very quickly and are rough approximations of a product, while others can take longer and be very good approximations of a product. Here is a description of a few types of prototypes.

- Rough Mock-up: Created by taping, gluing, or otherwise holding/arranging existing objects together to explore or communicate an idea, a rough mock-up can be extremely useful when coupled with brainstorming activities. It usually takes seconds to minutes to create rough mock-ups.
- Cardboard Prototype: Created by cutting, folding, or otherwise manipulating cardboard or foam-core to build an approximation of a product, cardboard prototypes are an excellent way to explore the general sizes and spatial relationships of major product components. Typically, very useful cardboard prototypes can be constructed in less than an hour.
- Foam Looks-like Prototype: Cut, carved, or otherwise shaped from high density urethane foam, after painting these prototypes can approximate very well the final look and feel of a product. These prototypes can take hours to days to create. Typically a very useful prototype can be constructed in a day.
- Rapid Prototype: Created layer-by-layer on a rapid prototyping machine that uses CAD data, these prototypes are very close to the final product geometry. If CAD models are available, very good approximations of products can be made within hours. Rapid prototyped parts can also be used as casting molds.
- Machined Works-like Prototype: Machined on conventional or CNC machining equipment, these prototypes often provide a robust platform for testing product functionality. They can be made with or without CAD data. These prototypes can take hours to days to create. If outsourcing the work, it will likely take longer than a week to get finished parts.
- Soft-tooling Prototype: Produced using production-style process (such as injection molding) on aluminum tools, rather than typical hardened tool-steel tools, these prototypes are nearly identical to the final product in esthetics and function. Soft tools are created in order to allow ready modification during initial production ramp-up. Soft tools may take several weeks to a few months, depending on complexity.

How to do it

The value of the prototype is in what design questions it answers and how well it does so. To that end, a plan should be made for each prototype[1] pursued. The following planning steps may prove useful:

1. Establish a purpose: Will the prototype be used for design learning? If so, what question will the prototype help answer? Will the prototype be used to communicate something? If so, what is it designed to communicate?

Further Reading

Stanford d.school's "Open vs. Closed Prototypes" which focuses on getting feedback from prototypes, available at: dschool.stanford.edu/blog/2011/02/28/open-vs-closed-prototypes/.

"Prototype to Test" which focuses on using prototypes to answer specific questions, available at dschool.stanford.edu/wp-content/themes/dschool/method-cards/prototype-to-test.pdf.

[1] It may not be necessary to have this plan for rough mock-ups.

2. Choose a target approximation level: Will the prototype be a cardboard prototype, a soft-tooling prototype, or something in between? As long as the prototype fulfills its purpose, the kind of prototype selected should be the one that costs the least in terms of time and money.

3. Create a prototyping schedule: The people who design the prototype are not always the people who make or test the prototype. To that end, it is useful to make sure all parties involved are aware of the required schedule. The schedule should be consistent with the purpose and the level of approximation.

4. Develop an experimental plan: What will be done with the prototype once it is in hand? Developing an experimental plan helps the team understand how many prototype units are needed and what tests should occur in what order. Generally, photographing and measuring the prototypes are the first action to be taken. Always do destructive testing after non-destructive testing (yes, teams do forget this from time to time!)

See also

Development Reference: CAD Modeling (11.6).

Section 6.6 of this book.

Product development prototypes for cabinet maker's impact driver. The swivel head allows the driver to get into tight spots. (a) Cardboard. (b) Wood. (c) Foam. (d-e) Fused deposition modeling rapid prototypes.

11.51 Quality Function Deployment

Use the voice of the customer to improve your design

Quality function deployment (QFD) is a formal method for ensuring that market requirements (called the voice of the customer or VOC) are fully considered at all levels of product development and production. The chart that is used to track the VOC is called the house of quality.

Further Reading

Chan LK, Wu ML (2002) Quality function deployment: A literature review. Eur J Oper Res 143:463-498.

Key methods

QFD uses a matrix for translating customer needs from one level to another. The top-level needs are placed on the left-hand side of the matrix and label the rows. They are called "Whats" because they describe what is to be achieved by the development project.

The columns of the matrix are called "Hows" because they describe how achievement of the Whats is to be measured or ensured. At the intersection of the rows and columns a symbol is placed indicating how the What and the How interact.

Interactions include three levels: weak, medium, and strong. Numerically, the three levels should form a geometric progression; the most common values are 1, 3, and 9, respectively.

An importance rating is given for each of the Whats. This rating captures the importance of the what to the customer.

The importance for each of the Hows is calculated from the importance of the Whats and the interaction ratings:

$$H_k = \sum_{j=1,N_w} I_{jk} W_j \qquad (11.11)$$

where H_k is the importance of the kth How, W_j is the importance of the jth What, N_w is the number of Whats, and I_{jk} is the interaction between the jth What and the kth How.

A triangular "roof" is placed above the Hows and is used to show interactions between the Hows. Interactions between Hows can be positive, meaning that achieving one How helps achieve the other How; or they can be negative, meaning that achieving one How hinders achieving the other How.

Competitive benchmarking information can also be placed on the QFD matrix. Market assessment of competitive products is placed at the right-hand side of the matrix, with an evaluation for each of the Whats. This portion of the matrix is called the How Wells, because it shows how well the market needs are met.

Technical assessment of competitive products is placed at the bottom of the matrix, with the performance of the product relative to the How indicated. This section of the matrix is called the How Muches because it is information on how much of the Hows should be provided.

Judgment is an important part of QFD application. The matrices can become so big that it is nearly impossible to finish them, or they can be so small that they are trivial and provide no benefit.

See also

Development Reference: Requirements Matrix (11.55), Six Sigma (11.62).

A sample of the house of quality.

Four levels of QFD matrices are often created. The top-level matrix relates the market needs to the design specifications. The second level relates the design specifications to design parameters. The third level relates the design parameters to part characteristics. The fourth level relates part characteristics to production system parameters. By means of these matrices, it is clear how day-to-day operations of the production system affect the quality of the delivered product.

Quality Function Deployment

11.52 Rapid Prototyping

Automatically build prototypes from CAD models

Rapid prototyping is a layer-by-layer additive manufacturing method that can be used to quickly create prototypes from CAD data. Rapid prototyping is a relatively inexpensive way to quickly obtain a physical object from the product definition.

Rapid prototyping is also known as 3D printing, solid freeform fabrication, and additive manufacturing.

Types of materials used

Rapid prototyping can be used with polymers (liquid, powder, filament), metals (powder, thin sheets), and paper. The resulting properties are generally inferior to the final product, but the geometry is nearly identical.

Types of machines

Rapid prototyping machines can be classified as *deposition machines* or *consolidation machines*.

Deposition machines carefully place material and fuse it together to build up geometry. A deposition machine has a computer controlled (x and y direction) head that extrudes material onto a build platform that progressively lowers (z direction). Undercuts are built using support material that is relatively easy to remove after the build is complete. Machine resolution is often described as z-drop height, which is typically 0.1 mm.

Consolidation machines use a computer controlled laser to selectively fuse material already in a bed or bath. When the process begins, the build platform sits just below the surface of the bed or bath. A laser solidifies or sinters the material from a thin layer of the part. The build platform drops a small amount, and the laser builds the new layer on top of the previous layer.

Strategies

Rapid prototyping can be used to make *parts* or to make *tools* that make parts. For example, rapid prototyping can be used to make the back housing for a mobile phone, or rapid prototyping can be used to create a mold of the housing. Resin can then be poured into the mold to make a non-rapid prototyped part. Rapid prototyped parts can also be used to create a casting mold.

Because the layer-by-layer construction method results in a part that does not have equal strength in all directions, the build orientation is a critical decision.

How to do it

1. Create the CAD Model.
2. Convert the CAD Model to data exchange format (generally STL).
3. Slice the STL File into thin cross-sectional layers.
4. Layer-by-layer Construction.
5. Clean and post-process the part.

See also

Development Reference: Prototyping (11.50).

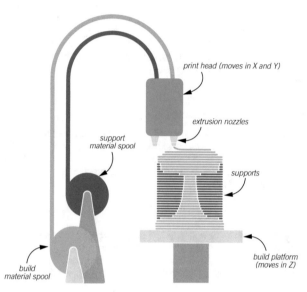

This deposition style machine is representative of Fused Deposition Modeling. The part is built up as filament is extruded onto a build platform that progressively lowers.

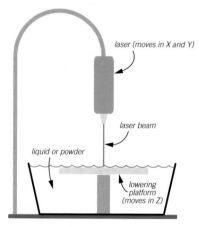

This consolidation machine, representative of the SLA machine, builds the part by solidifying a photosensitive polymer liquid in a bed.

Rapid Prototyping

CD

SSE

11.53 Recombination Table

Combine subconcepts to create hundreds of alternatives

Recombination tables are used to combine subfunction concepts together into full-function (full-system) concepts. One of the table's strengths is the large number of full-function concepts that can emerge. While not all emerging concepts will be feasible, the table makes it possible to explore the design space in a systematic and likely more thorough way. The table can also facilitate concept exploration activities as it helps the team visualize how well distributed across subfunctions the concept exploration has been.

Parts of the recombination table

The leftmost column of the table lists the subfunctions that a product needs to meet. These often come directly from a functional decomposition (see Decomposition). In each row of the table, corresponding to the subfunction, concepts are listed in the columns. It's not necessary that an equal number of subfunction concepts be generated for each subfunction.

Example

Consider a bicycle-like device for transporting humans. The required subfunctions are listed in the table. Now, concepts are generated for each subfunction. For example, the first subfunction listed is *power vehicle*. Various concepts for meeting this function include human power, internal combustion engine, electric motor, and human-electric hybrid.

Facilitated by the table, the team can now perform the recombination by simply picking a concept for one subsystem and combining it with one from all of the other subconcepts. The team can do this by intuitively picking concepts that seem compatible or otherwise worth exploring. Or the team can do this in an automated way, where every possible combination is made.

The maximum number of unique combinations is the number of concepts for subfunction one times the number of concepts for subfunction two, and so on. For this bicycle example, there are a total of 192 full-function (full-system) concepts.

See also

Development Reference: Decomposition (11.14).

Functions	Subfunction Concepts			
Power Vehicle	*Human Powered*	*Gas Motor*	*Electric Motor*	*Human-electric Hybrid*
Support Rider	*Small Seat*	*Banana Seat*	*Bucket Seat*	*Suspended Hammock*
Contact Ground	*1 Wheel*	*2 Wheels*	*3 Wheels*	*4 Wheels*
Aesthetic	*Orange*	*Yellow*	*Blue*	
...				

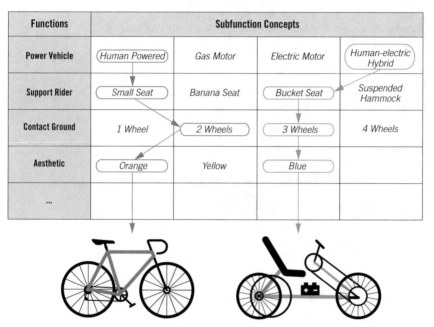

Recombination Table for Bicycle Example.

11.54 Requirements Hierarchy

Develop a comprehensive list of market requirements

A requirements hierarchy is a list of market requirements that is organized into groups of related requirements, with each group given a descriptive title. The requirements hierarchy allows the design team to capture market requirements in a way that will be very useful for guiding product development.

How to do it

The first step in developing a requirements hierarchy is to obtain input from potential users. This can be done in several ways, including surveys, focus groups, interviews, direct observation, and more. The user input should always be captured in the user's own words to the extent possible.

Next, the user statements are rewritten as *product-focused requirement statements*, as described in Product-Focused Requirement Statements (11.48.)

The requirements statements are then organized. The set of product-focused requirement statements needs further work to be of most use in product development. There may be hundreds of requirement statements in the list, which is too many for the design team to effectively consider. Some requirements statements will be very specific, while others will be general. For most cases, the requirements hierarchy will be most helpful if it contains between five and twenty requirements, all of which are at approximately the same level of generality.

The requirements statements are organized by grouping related requirements statements, eliminating duplicates, and writing new requirement statements that effectively summarize all of the statements in the group. These primary requirement statements have all the attributes of the product-focused requirement statements, except that they summarize the group of related requirements, rather than restating each of the requirements in the group.

The primary requirements are then placed in the requirements matrix as the market requirements.

See also

Development Reference: Product-Focused Requirement Statements (11.48), Requirements Matrix (11.55).

Capstone students were asked to develop a piece of playground equipment that would generate electricity while students played on it. The electricity would be used to charge batteries that would power LED lanterns to provide light in the school classrooms as well as for doing home-work. They developed a merry-go-round that drove a generator.

Primary requirement statement	Original requirement statements
Equipment is safe to use	Equipment feels safe Equipment prevents user from falling off or sliding out Equipment feels like it will not break Equipment prevents blistering of hands Equipment prevents injuries to users
Equipment reliably provides electricity	Equipment parts prevent slipping Equipment generates sufficient current and voltage to light one classroom for an hour per hour of play Equipment fits on or into normal Ghanaian school ground
Equipment works well in Ghana	Equipment withstands the heavy rain and dust storms of Ghana Equipment withstands many children using it at the same time Equipment can be manufactured in Ghana
Equipment challenges the users	Equipment allows "push it to the limit" Equipment provides a competition that can last a long time Equipment is a proving ground (prove to others) Equipment makes the user feel powerful Equipment gives the user an opportunity to show off Equipment allows the user to impress him- or herself with strength Equipment builds confidence in self (prove to self) Equipment provides an accomplishable challenge
Equipment provides fun motion	Equipment provides a long time of motion Equipment is big and fast Equipment allows the user to move Equipment moves fast
Equipment is fun and exciting	Equipment looks fun to play with Equipment is new and exciting Equipment resembles something exciting to children in Ghana Equipment provides a thrill for the user Equipment allows friends to gather Equipment holds multiple children safely
Equipment may be used creatively	Equipment may be used creatively

Based on statements from market representatives, a set of product-focused requirement statements was developed and organized into a requirement hierarchy. The primary requirements were used as the market requirements in the requirements matrix.

Requirements Hierarchy

OD

CD

11.55 Requirements Matrix

Capture all information about meeting the market requirements in one place

The requirements matrix is a convenient place to store the requirements in a way that is clear, unambiguous, and easily transferable. More importantly, however, it is a powerful tool for translating broad market requirements into specific performance measures that are generally quantitative. To be useful, it is important that the requirements matrix be updated frequently and used to track development progress.

Sections of the requirements matrix

There are six main sections (A–F) to the requirements matrix. Each is discussed briefly below. The inset image in the example on the facing page indicates the locations of A–F.

Market requirements (A)

These are the top-level requirements for the product to be successful in the market. They are often identified/clarified by interviewing clients, end users, and experts, and by benchmarking and other forms of research. Market requirements are statements about *what* the market wants in the product, and are often subjective.

Optionally, the relative importance of each requirement can be assigned. More important wants have larger importance numbers. The scale used to measure importance is generally relatively coarse.

Often, the market requirements are the primary requirements from a requirements hierarchy.

Performance measures and units (B)

These are specific characteristics of a product that can be measured to determine how well a product meets the market requirements. They can be thought of as *how* the market requirement will be evaluated or measured. It is generally preferable to have objective performance measures. When subjective performance measures are necessary, they should be unambiguously evaluated. In general, there are multiple performance measures for each market requirement. Likewise a performance measure can often apply to multiple market requirements. The units of the performance measure are listed next to the measures.

Optionally, the relative importance of each performance measure can be assigned. Higher importance is indicated by larger importance numbers. In QFD, the importance of the performance measures is calculated from the importance of the market requirements and the requirement–measure relationships, but this is not essential.

Requirement–measure correlations (C)

This portion of the matrix captures the correlations between market requirements and performance measures. In the figure, a dot in a cell shows that the requirement in the row is at least partially evaluated by the performance measure in the column.

Ideal values (D)

This section indicates the values that the market would like to have for each of the performance measures, in the absence of any trade-offs. There are three kinds of possible ideal values.

The *Lower Acceptable Limit* indicates the smallest value for the performance measure that the market deems acceptable.

The *Upper Acceptable Limit* indicates the largest value for the performance measure that the market deems acceptable.

The *Ideal* value indicates the value that the market would prefer if no trade-offs are required.

Some performance measures will have only one acceptable limit.

The ideal values can be placed in the requirements matrix during the opportunity development stage.

Together, regions A, B, C, and D comprise the market opportunity. The market opportunity is created during opportunity development.

Real values (E)

This section indicates the values that are desired, predicted, and measured for the product. Although it would seem that the target values should be the same as the ideal values, this is not true. It may be impossible to simultaneously achieve all the ideal values given technological limits of the selected product concept.

The target values represent the values the team has decided to pursue given the trade-offs inherent in the selected product concept. These values cannot be added to the requirements matrix until the product concept is selected in Concept Development.

During Subsystem Engineering and System Refinement, the design is finalized. Models that relate the design to the performance measures are developed. When the models are applied to the design in accordance with the tests, predicted values of the performance measures are obtained.

The predicted values will change during product development as the design and tests evolve, and as different models are created. Throughout the development process, the most current predicted performance values should be listed in the requirements matrix.

As prototypes are developed from the design and tested according to the tests, measured values of the performance measures are obtained. As the product information and tests evolve, the measured values will change. The most current measured values should be listed in the requirements matrix.

Market response (F)

This represents the market's rating of the product relative to the market requirements. This information is available only at the end of the system refinement stage when complete prototypes are available for the market to evaluate.

If desired, the market rating of competitive products can be listed in this section as well.

Using the requirements matrix

The matrix is created in the opportunity development stage to capture market requirements and translate them into performance measures that will guide the

Requirements Matrix

development efforts. It is most easily used by first listing market requirements in section A. Then, for each market requirement, the team decides what performance measures would best represent the market requirement. These are placed in section B. The team seeks unambiguous objective performance measures. When objective measures are not possible, the team creates unambiguous subjective performance measures. This is continued for each market requirement. Next, the team indicates the relationships in section C. Finally, through dialogue with others, benchmarking, and other research, the team establishes the ideal values in section D.

During concept development, after a concept is chosen, the team reconsiders the performance measures and the trade-offs that exist between them for the chosen product concept. This leads to the selection of target values for each of the performance measures, which are placed in section E.

During concept development, as subsystems are identified, requirements matrices are developed for each of the subsystems. Subsystem requirements matrices relate system performance measures (the Whats of the subsystem matrix) to subsystem performance measures (the Hows of the subsystem matrix).

During any stage of development, the predicted and measured values are placed in section E of the matrix. The predicted and measured values are compared with the target values during the approval at the end of various stages.

At the end of the system refinement stage, prototypes of the product can be evaluated by the market representative and the results of the evaluation placed in section F.

The matrix is a valuable way to capture product performance in an unambiguous and transferable way. Like most product development tools, however, the value goes well beyond this. When used wholeheartedly, the understanding gained about the product design makes the time spent preparing the matrix well worth it.

See also

Development Reference: Product-Focused Requirement Statements (11.48), Quality Function Deployment (11.51), Requirements Hierarchy (11.54).

Requirements Matrix

Product: DRILL
Subsystem: N/A

Performance Measures (Units)

#	Performance Measure	Units
1	Maximum borehole depth	ft
2	Time required to cut through 6 inches of rock	min
3	Downward drilling force	lbs
4	Torque applied to drill bit	ft-lbs
5	Compatable with X% existing drill bits	%
6	Water pressure down the pipe	psi
7	Percentage of water that leaks through sides	%
8	Percentage volume of cuttings removed	%
9	Depth cut per 8 hours of drilling	ft
10	Number of required people	people
11	Weight of heaviest subassembly	lbs
12	Longest dimension (l,w,h) of biggest subassembly	in
13	Percentage of drill manufacturable in Tanzania	%
14	Cost to produce 1 drill after development	USD
15	Time required to learn how to operate	hr
16	Height of hand operated parts	ft
17	Feels comfortable	n/a
18	The Drill is attractive	n/a
19	The Drill interests investors	n/a

Market Requirements (Whats) — relationships to Performance Measures

#	Market Requirements (Whats)	Importance	Related PMs (●)	Market Response
1	The Drill reaches potable water beyond 100 ft	9	1, 2	Excellent
2	The Drill cuts through rock	1	2, 3, 4	Acceptable
3	The Drill uses existing drill bits	3	5	Excellent
4	The Drill seals borehole sides to prevent cave in	3	6, 7	Very Good
5	The Drill removes cuttings from the borehole	3	6, 8	Excellent
6	The Drill works at an efficient speed	3	9	Excellent
7	The Drill uses only manual labor to function	9	3, 4, 8	Acceptable
8	The Drill is affordable	9	13, 14, 18	Very Good
9	The Drill requires simple training to operate	9	15	Excellent
10	The Drill is portable	9	11, 12	Acceptable
11	The Drill is comfortable to operate	1	16, 17	Very Good
12	The Drill is attractive	1	18	Excellent
13	The Drill attracts investors	3	18, 19	Excellent

Target / Measured values by Performance Measure

PM #	Importance	Upper Acceptable	Ideal	Lower Acceptable	Target	Predicted	Measured
1	9	–	250	100	220	–	140
2	10	60	45	–	60	–	480
3	10	–	3,000	500	200	–	1313
4	10	–	400	200	300	–	Not Tested
5	3	–	100	90	95	–	100
6	6	–	113	50	113	–	65
7	3	5	0	–	5	–	Not Tested
8	3	–	100	95	95	–	Not Tested
9	3	12	36	4	36	–	182
10	9	–	3	–	4	–	4
11	9	400	50	–	200	–	332
12	9	96	48	–	96	–	84
13	9	–	100	85	90	–	66
14	9	5,000	1,000	–	1,500	–	1,600
15	9	20	4	–	8	–	4
16	1	5	3.5	2	3.5	–	3.5
17	1	–	Drill can be operated continuously without the need to rest. Does not require awkward movements.	Drill can be operated with occasional rest, and requires awkward movements that leave the user sore.	Ideal	–	Marginal Measured
18	4	–	Drill has a professional look. People are interested in looking at it.	Drill looks like a piece of machinery for drilling holes.	Ideal	–	Ideal Measured
19	12	–	Drill captures media attention. Investor s are proactive in contributing. Drill has iconic look.	The drill, when explained to investors, is something they want to invest in.	Ideal	–	Ideal Measured

Correlation roof legend:

	B	
A	C	F
	D	
	E	

A simplified requirements matrix for the human-powered water well drill described in Appendix B of the book. Not all of the market requirements and performance measures have been listed for reasons of space.

Requirements Matrix

Revision Control

11.56 Revision Control

Track the evolution of product development artifacts

During product development the requirements, tests, and design evolve. This evolution is captured in discrete steps through the creation of product development artifacts. As the information evolves, the artifacts will change. To keep track of these changes, the artifacts must be placed under revision control.

Principles of revision control

- All product development artifacts should be placed under revision control. This means that, in addition to the design, the requirements and the tests should be controlled.

- Revision control is required for effective coordination between and within teams, because it allows all parties to be sure they are working with the same information.

- A design artifact should be given a unique revision identifier when changes have been made and the artifact is to be shared beyond its creator. If a CAD model is saved every thirty minutes, it need not be given a new revision identifier with each save. However, if a drawing made from the model is sent to be checked, both the CAD model and the drawing should have a revision number, even if the last change was only ten minutes ago.

- While an artifact is being created, before formal approval, the revision identifier should identify it as unreleased, generally by having a different numbering system. The revision process for unreleased artifacts is less formal than that for released artifacts. However, pre-release versions should still be tracked.

- After formal approval, revision to artifacts requires the use of an engineering change order (ECO) process to ensure that all who would be affected by the change are made aware of it.

- All revisions of released artifacts should be archived so that they are available for review at any time. It may be necessary to refer to prior revisions in order to provide support for older versions of the product.

- Small changes in a design (often defined as those that will not be visible to the end user or customer) can sometimes have a minor revision number and a less rigorous ECO process.

How to do it

The following tips can help develop and maintain a revision control system.

- Use a detailed Bill of Materials (BOM) as an index to all of the components in the design. The BOM itself should be under revision control. Each component listed in the BOM should also have a revision number.

- Keep a list of requirement and test artifacts as an index to all of the requirements and test information. The list of artifacts should be under revision control. Each of the listed artifacts should have a revision number.

- When a revision number is assigned, save a copy of any computer file with the revision number as part of the name. Revision control software or product data management software can help with this task.

- Create a formal definition of the change process for released artifacts. Ensure that the defined change process is followed.

OD

CD

SSE

SR

PR

PRR

- Develop a culture of revision control. Revision control takes work, and the work of managing revisions is usually not exciting for designers. If the culture of revision control is not established, it's likely that revision control won't happen. With strong cultural expectations, revision control can become a regular and valued part of the process.

Applicability

Revision control is used through all stages of development on all product development artifacts.

See also

Development Reference: Bill of Materials (11.3), Drawings (11.23), Engineering Change Order (ECO) (11.24).

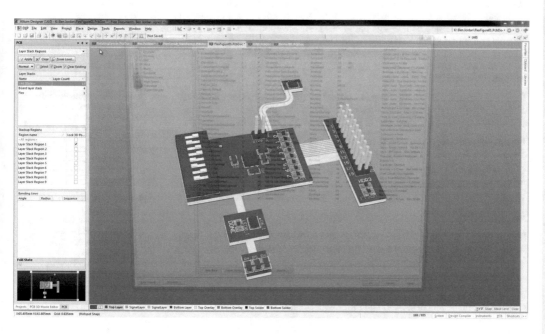

Revision control software can simplify the creation and management of design revisions.

Revision Control

11.57 Robust Design

Design your products to work all the time, in all conditions

Robust design is a method used to ensure that a product will function well even in non-ideal operating conditions and with the normal variability in manufacturing processes. It seeks to maximize signal-to-noise ratios, where signal is the nominal performance and noise is the variation in performance due to deviations from ideality.

Key methods

Robust design aims to minimize the effects of variation on the performance of the product that is made. It includes efforts to improve the design as well as efforts to improve the production process. Robust design was popularized in the 1980s by Gen'ichi Taguchi.

Taguchi developed special experimental designs that include both design parameters (called the inner array) and environmental or noise factors (called the outer array). By varying both design parameters and environmental factors, both the nominal performance of the design and the variation due to environmental noise can be estimated. It is then possible to identify parameter settings that not only have high absolute performance, but also have minimal variation due to the environmental variables.

To find robust operating points, Taguchi seeks to maximize signal-to-noise ratios, where signal is the mean performance of the design at a given set of parameters, and noise is the standard deviation of the design at the same set of parameters.

Taguchi experiments aim to create response surfaces of S/N ratio, so that a location of maximum S/N ratio can be found.

Taguchi provides a variety of different experimental plans, with varying numbers of design parameters and noise factors.

Applicability

Robust design is generally applicable in subsystem development and later stages. Robust design thinking should be a consistent part of product development in order to produce high-quality products.

Some American statisticians believe that the specific experimental designs Taguchi promotes are less effective than other designs would be. However, the idea of considering the effects of variation on the design is supported by these same statisticians.

See also

Development Reference: Design of Experiments (11.18), Six Sigma (11.62), Sensitivity Analysis (11.61).

Further Reading

Taguchi G, Clausing D (1990) Robust quality. Harvard Bus Rev, Jan-Feb:65-75. An article written by Taguchi for U.S. management.

Phadke MS (1995) Quality engineering using robust design Prentice Hall, Upper Saddle River New Jersey. A book written by a student of Taguchi that covers Robust Design thoroughly.

Phadke MS Introduction to robust design (Taguchi method) is available at www.isixsigma.com/methodology/robust-design-taguchi-method/introduction-robust-design-taguchi-method/

Box G (1988) Signal-to-noise ratios, performance criteria, and transformations. Technometrics 30(1):1-17. This paper claims that traditional DOE techniques are better than Taguchi methods for creating robust designs.

Hunter RG, Sutherland JW, Devor RE (1989) A methodology for robust design using models for mean, variance, and loss. Proc AMSE WAM, San Francisco:25-42. This paper compares Taguchi experimental designs with other design of experiments approaches.

Robust Design

In robust design, products are tested under conditions that are far from ideal to determine how well they will perform in adverse conditions.

11.58 SCAMPER

Change existing products to create new concepts

SCAMPER is an acronym that captures techniques sometimes used by companies to expand their product offering by rethinking one or more of their existing products.

Further Reading

Thought questions and keywords to help with each technique can be found at litemind.com/scamper.

How to do it

Generally SCAMPER can aid brainstorming. Formally as a group, or individually in one's head, the points of the acronym can be used to ask the following questions:

1. Substitute: What if a different material, process, person, or place were substituted for the existing ones? For example, could the handles on a pair of scissors be made with a different material? Would doing so produce a valuable product variant?
2. Combine: Can we combine units, purposes, or ideas? For example, can a standard pair of scissors be combined with another useful tool such as a bottle opener?
3. Adapt: What adaptations could be made to the idea or product to make it useful in a similar context. For example, what adaptation would be needed to make a standard pair of scissors useful for cutting sheet metal?
4. Modify/Magnify: Can the form, size, color, meaning, or motion be modified? For example, we could modify the form of the scissor blades that produce decorative cuts.
5. Put to other uses: Are there new ways to use the product without modifying it? The scissors could be used as a box cutter, for example.
6. Eliminate: Can a part, function, or person be eliminated? Could the person be removed as the actuator, for example? This could result in a machine that performs the actuation, allowing the person to control only the direction and speed of cut.
7. Rearrange/Reverse: Can a different layout or sequence be used? The pin joint could be moved to a different location on the scissors, for example.

See also

Development Reference: Brainstorming (11.5).

SCAMPER

285

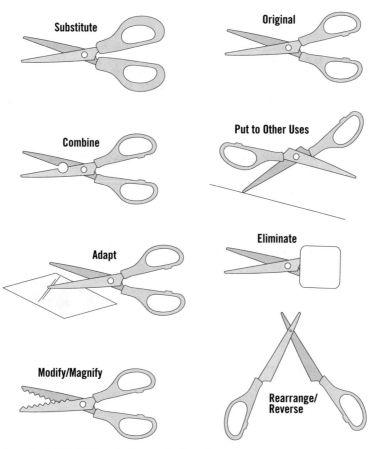

Example of SCAMPER method as applied to scissors.

SCAMPER

CD

11.59 Scoring Matrix

SSE

Evaluate concepts with greater resolution

A scoring matrix is a tool used to evaluate concepts at a finer resolution than the scoring matrix in order to provide guidance for selection of a superior concept. Generally, a scoring matrix is used when there are fewer than ten concepts to evaluate. The matrix is designed to evaluate each concept relative to weighted market requirements (or any other design criterion).

How to do it

Construct the matrix: List the market requirements in column one. Assign weighting factors w_i for each concept and list them in column two. These weights most often range between zero and one, and sum to one. List the concepts at the top of the remaining columns of the matrix.

Rate the concepts: Moving row by row, one requirement at a time, rate each concept as compared to the reference concept. The ratings (r_{ij}) typically take on numerical values from the five-point rating scale as shown in the table. The five-point rating scale allows more resolution than the three-point scale used in a screening matrix.

One concept can be selected as a reference concept (as shown in the example), or each requirement can have a different reference concept.

Two important objectives are met by moving row by row during rating. First, it is much easier to be consistent in the evaluation of a particular market requirement if all the concepts are evaluated for that market requirement before moving on to other requirements. Second, completing an entire row before moving to the next prevents gamesmanship in artificially raising or lowering scores for specific concepts. Gamesmanship can also be avoided by requiring the team to state a rational reason for the rating given each concept.

Calculate the score for each concept: The weighted score for each concept is calculated as the sum of the weighted ratings or

$$S_j = \sum w_i r_{ij}$$

The use of non-uniform weighting factors allows more important requirements to have greater influence on the concept score.

Use the weighted score estimate to concept quality: The weighted score is used as a prediction of how well a product based on a particular concept would meet the market requirements. Concepts with higher scores are considered better than those with lower scores.

The best concept is not necessarily the one that has the highest score from the scoring matrix. In fact, it is possible that several concepts will have similar scores, or that there are other criteria that are important to the team but not listed in the matrix. In this sense, the matrix does not make the decision. It simply provides information to help the designer make a decision.

An important, often overlooked benefit of using a scoring matrix is the opportunity to combine concepts that are strong in one area with others that are strong in different areas. The same approach can be used to strengthen concepts that are particularly weak. In this way, the scoring matrix is more than an evaluation tool – it facilitates systematic concept development and evolution.

See also

Development Reference: Controlled Convergence (11.10), Screening Matrix (11.60).

	Weight	Concept 1	Concept 2	Reference	...	Concept M
Requirement 1	w_1	r_{11}	r_{12}	r_{1REF}		r_{1M}
Requirement 2	w_2	r_{21}	r_{22}	r_{2REF}		r_{2M}
Requirement 3	w_3	r_{31}	r_{32}	r_{3REF}		r_{3M}
...						
Requirement N	w_N	r_{N1}	r_{N2}	$r_{N,REF}$		$r_{N,M}$
Weighted Score		S_1	S_2	S_3		S_4

Concept Scoring Matrix.

Rating	Description
1	*Much Worse Than Reference*
2	*Worse Than Reference*
3	*Same As Reference*
4	*Better Than Reference*
5	*Much Better Than Reference*

Typical 5-point scale for rating concepts.

11.60 Screening Matrix

Quickly evaluate a relatively large number of concepts

Concept screening is a concept evaluation method that takes a coarse look at various candidate concepts and rates them relative to a reference concept (sometimes called benchmark design or concept). The method is conveniently carried out with the help of a concept screening matrix.

A screening matrix is a quick and effective way to reduce the set of concepts from roughly 20 to less than 10 and to improve/combine concepts during the process. The matrix is designed to evaluate each concept relative to each market requirement (or any other design criterion).

How to do it

To construct the screening matrix, list the market requirements in column 1. List the concepts at the top of the remaining columns. For each requirement (or criterion) rate each concept as better than (+), same as (=), or worse than (−) the reference concept. Count the +'s, ='s, and −'s in each column. Calculate a net score by subtracting the number of −'s from the number of +'s . Consider the number of +'s, ='s, and −'s to identify the overall strengths and weaknesses of the concepts.

When making a rating, the team should reach consensus in order to give a rating of + or −. If the team cannot agree on one of these ratings, the default rating of = is used.

The choice of the reference concept is up to the team. It is often the lowest-risk or least novel concept in the set.

Rate all the concepts for one requirement before moving on to the next requirement, rather than rating a single concept for all requirements. Two important objectives are met by moving row by row in this way. First, it is much easier to be consistent in the evaluation of a particular market requirement if all the concepts are evaluated for that market requirement before moving on to other requirements. Second, completing an entire row before moving to the next prevents gamesmanship in artificially raising or lowering scores for specific concepts.

After rating all concepts for all requirements, sum the ratings and calculate the net score as described above. It's important to recognize that the best concept is not necessarily the one that has the highest score from the screening matrix. In fact, it is possible that several concepts will have equal scores, or that there are other criteria that are important to the team, but not listed in the matrix. In this sense, the matrix does not make the decision. It simply provides information that allows the human to make a decision.

An important, often overlooked benefit of using a screening matrix is the opportunity to combine concepts that are strong in one area with others that are strong in different areas. The same approach can be used to strengthen concepts that are particularly weak. In this way, the screening matrix is more than an evaluation tool – it facilitates systematic concept development and evolution.

See also

Development Reference: Controlled Convergence (11.10), Scoring Matrix (11.59).

	Concept 1	Concept 2	Reference	...	Concept M
Requirement 1	+	+	=		+
Requirement 2	+	−	=		+
Requirement 3	=	−	=		=
...					
Requirement N	−	−	=		=
Number of +'s	2	1	0		2
Number of ='s	1	0	4		2
Number of −'s	1	3	0		0
Net Score	1	−2	0		2
Improve/Combine?	*Improve*	*Combine*	*Combine*		*None*

Concept Screening Matrix.

11.61 Sensitivity Analysis

Understand how small changes will affect performance

Sensitivity analysis is a technique used to understand the importance of each of the parameters in a model or product to the performance of the product. It provides understanding about how changes in individual components or features affect performance, and is therefore a tool for rational selection of component tolerances.

Key methods

The first step in sensitivity analysis is to determine the performance for which we want to do the analysis. The performance can be modeled analytically or it can be measured experimentally. However, we must have some means of obtaining a quantitative measure of performance

The second step is to make a first-order Taylor series approximation of the performance. We do this by varying each of the parameters that affects the performance, one-by-one. This will require one (if we are content to use a single-sided derivative approximation) or two (if we wish to use a double-sided derivative approximation) evaluations of the performance for each parameter.

The third step is to estimate the resolution of each of the parameters. Formally, the resolution is the standard deviation of the parameter. Informally, it may be the specified tolerance on the parameter.

The final step is to calculate the sensitivity for each individual parameter:

$$S_i = \frac{\Delta Y}{\Delta P_i} \delta_i \qquad (11.12)$$

where S_i is the sensitivity for parameter P_i, Y is the performance measure being studied, $\frac{\Delta Y}{\Delta P_i}$ is the numerical derivative of the performance with respect to parameter P_i, and δ_i is the resolution of parameter P_i.

Having measured sensitivities for each of the parameters, we can now see which parameters have the greatest effect on the performance, and make plans to improve the resolution or adjust values of the parameters with the largest sensitivities.

Sensitivity analysis is similar to *uncertainty analysis*, but uncertainty analysis works with closed-form solutions where partial derivatives can be calculated analytically, and sensitivity analysis works with numerical solutions where partial derivatives must be calculated numerically.

Applicability

Sensitivity analysis is most often used in the subsystem engineering and system refinement stages of development.

See also

Development Reference: Design of Experiments (11.18), Uncertainty Analysis (11.68).

Further Reading

Saltelli A, Ratto M, Andres T, Campolongo F, Cariboni J, Gatelli D, Saisana M, Tarantola S (2008) Global Sensitivity Analysis. The Primer, John Wiley & Sons, New York.

Wikipedia has an excellent entry on sensitivity analysis: http://en.wikipedia.org/wiki/Sensitivity_analysis

SSE

SR

Sensitivity Analysis

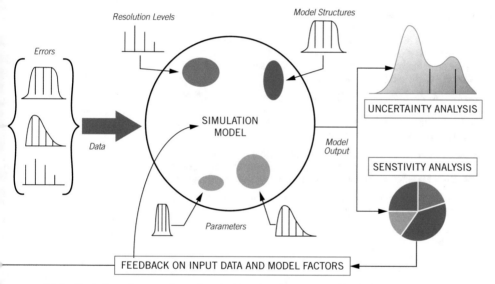

Resolution Levels

Model Structures

Errors

SIMULATION MODEL

Data

UNCERTAINTY ANALYSIS

Model Output

SENSTIVITY ANALYSIS

Parameters

FEEDBACK ON INPUT DATA AND MODEL FACTORS

How uncertainty affects the outcomes of experiments. Based on image created by Andrea Saltelli.

Sensitivity Analysis

11.62 Six Sigma

Improve quality by designing for manufacturing variations

Six Sigma (6σ) is a set of methods to ensure that designs are robust relative to manufacturing variations. It is used to adjust both product and process designs to achieve high quality and low variation.

Key methods

Because Six Sigma is a collection of methods, there is no space to describe the methods here. Instead, we provide a fundamental philosophical basis for Six Sigma, and encourage the interested reader to review the literature for detailed instructions on the methods.

Six Sigma started at Motorola, where it was recognized that traditional quality limits would not suffice in manufacturing environments like electronics that have thousands of components. Standard quality limits of plus or minus 3σ (where σ is the process standard deviation) lead to approximately 3 defective parts per thousand. If there are a thousand components in the assembly, this would mean that each assembly would have three defective components.

If instead, quality limits are specified at plus or minus 6σ, even if a 1.5σ drift occurs, there will only be 7 defective components per million, so for a one-thousand-component assembly, there will be only 7 assemblies containing a defective component per every million assemblies.

Motorola developed a plan for reducing variability in production processes and increasing the robustness of designs. This combined reduction in standard deviation and increase in tolerance limits could lead to plus or minus 6σ limits for the process.

Six Sigma gained widespread popularity when Jack Welch applied it throughout GE, which led to large increases in profitability.

Six Sigma has developed into a full-blown quality system, with training and certification available.

Applicability

Six Sigma techniques are often applied during system refinement, producibility refinement, and post-release refinement.

See also

Development Reference: Design of Experiments (11.18), Robust Design (11.57), Sensitivity Analysis (11.61).

Further Reading

Tennant G (2001) Sis sigma: SPC and TQM in manufacturing and services. Gower Publishing, Ltd.

Harry MJ, Mann PS, De Hodgins OC, Hulbert RL, Lacke CJ (2011) Practitioner's guide to statistics and lean six sigma for process improvements. John Wiley & Sons, New York.

SR

PR

PRR

Six Sigma

11.63 Sketching

Communicate design ideas with high-quality sketches

Sketching is a powerful part of product development. It is a skill everyone could use, and everyone could work on improving. We sketch to facilitate concept exploration, and we sketch to communicate our ideas.

Those trained in the art can produce sketches such as those on the facing page. Such sketches are very useful in sharing ideas and helping others embrace them.

Sketching tips

Those of us not trained in the art might find these sketching tips useful.

- Lose drawing inhibition: Forget about what others will think of how good or bad your sketch is.
- Practice: Good sketchers practice. We know some who do sketch warm ups; they draw pages of straight lines, pages of circles, pages of ellipses, and so on.
- Sketch with your arm (not your finger or wrist) this will produce nice, confident lines. Not doing this tip will result in chicken scratch (wiggly lines). See sub-figure b in the facing page.
- Complete line connections: Incomplete connections are cognitively unpleasing (sub-figure c).
- Build from basic shapes: Almost everything can be constructed from lines and ellipses. Use lines to make cubes. Use lines and ellipses to make cylinders (sub-figure d).
- Try perspective view using vanishing points (sub-figure f): This is more realistic than isometric figures (sub-figure e).
- Try shading and shadows (sub-figure g).

See also

Development Reference: CAD Modeling (11.6).

Further Reading

Tutorials for engineers can be found at web.mit.edu/2. 009/www/resources/ sketchingTutorials.html.

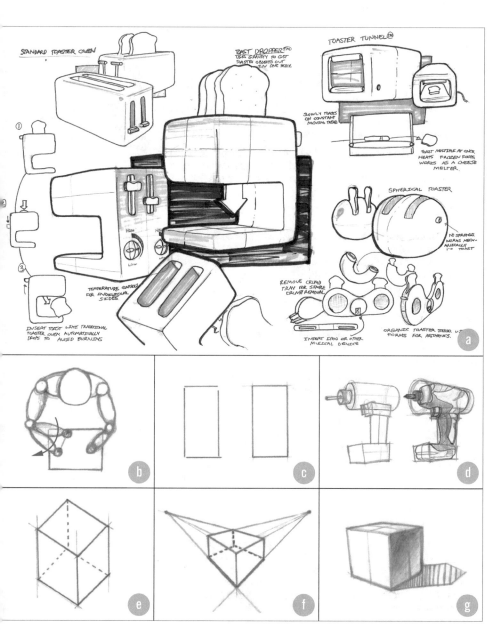

oncept sketches of a toaster (a). Sketches by Stephen Jensen, used with permission.

Sketching

11.64 Storyboards

Envision what end users will experience using your product

While developing a product, it is surprisingly easy to lose sight of *who* will interact with the product once it is on the market and *what* their experience with it will be.

Storyboards visually represent moments in the user's experience with the product. They are a powerful way to quickly convey a user-centered experience to project stakeholders, other team members, or potential end users. They help designers envision and ultimately plan for a meaningful experience for end users early in the development process. Storyboards help team members focus on what the end user will do with the product and how they will interact with it.

Because of their simple language and focus on the human experience, storyboards are particularly useful at assisting others – of different backgrounds, disciplines, or cultures – to understand and evaluate a problem or potential product solution.

Further Reading

van Boeijen A, Dallhuizen J, Zijlstra J, van der Schoor R (2014) Delft Design Guide BIS Publishers. pp. 152-153.

Hanington B, Martin B.(2012) Universal methods of design: 100 ways to research complex problems, develop innovative ideas, and design effective solutions. Rockport Publishers. pp. 170-171.

How to create storyboards

1. Develop a good understanding of who will use the product. Choose a subject (character) for the storyboard.
2. Choose one or more of the following to represent in a storyboard, and consider what the subject will experience when they:
   - Find out about the product
   - Decide whether they want the product or not
   - Acquire the product
   - Unpack, install, and/or learn how to use the product
   - Use and/or maintain the product
   - Dispose of or recycle the product
3. Tell the end user's story with the product. Do this in a graphical way, using minimal words. Consider the following visual elements:
   - Graphical medium (sketch, photo, computer line-art, etc.)
   - Number of panel (typically 12 or less)
   - Camera angle and zoom (to focus attention on one area)
   - Techniques to convey movement (arrows, wind, etc.)
   - Techniques to convey sound (onomatopoeia, sound waves, etc.)
   - Techniques to convey emotion (eyebrows, shoulders, head angle, etc.)
   - Use of dialogue and other words (speech bubbles, captions, etc.)

How to use storyboards to advance the design

Bring a storyboard to a team meeting and see how it changes the discussion from that of *product* to that of *how the product betters the user's life* or *solves a user problem*. Try using a storyboard to envision the user experience and brainstorm/refine market requirements. Use storyboards to convey concepts quickly to a non-technical audience and to get feedback. Use storyboards to promote discussion and build end-user empathy. Try using storyboards as a low-fidelity validation prototype.

See also

Development Reference: Personas and Locales (11.45), Sketching (11.63).

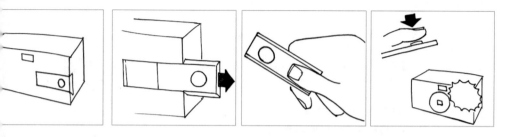

Simple hand sketched storyboard used to convey camera remote shutter button concept.

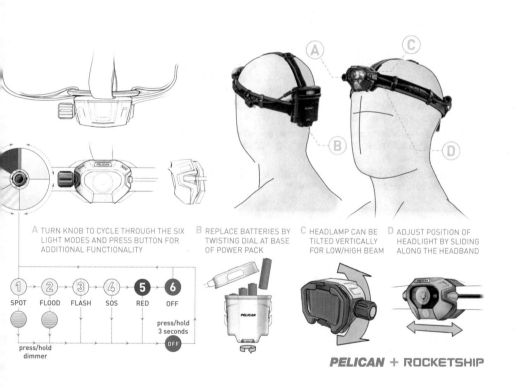

A TURN KNOB TO CYCLE THROUGH THE SIX LIGHT MODES AND PRESS BUTTON FOR ADDITIONAL FUNCTIONALITY

B REPLACE BATTERIES BY TWISTING DIAL AT BASE OF POWER PACK

C HEADLAMP CAN BE TILTED VERTICALLY FOR LOW/HIGH BEAM

D ADJUST POSITION OF HEADLIGHT BY SLIDING ALONG THE HEADBAND

1 SPOT 2 FLOOD 3 FLASH 4 SOS 5 RED 6 OFF

press/hold 3 seconds
OFF

press/hold dimmer

PELICAN + ROCKETSHIP

Detailed storyboard showing how to set, use, and adjust a headlamp.

11.65 Surveys

Obtain market information from many individuals

Surveys are written, in-person, telephone, or mail-based instruments used to better understand the preferences of the market.

Surveys are generally cheaper to conduct than focus groups, but are harder to get effective open-ended information from.

Guidelines for creating effective surveys

The following steps have been found useful when creating surveys:

- Define the purpose of the survey: A narrowly focused survey will be far more effective at obtaining useful information than a broadly focused survey. Defining the purpose is essential to narrow the focus.
- Define the target audience: Surveys should be tailored to a specific target audience. Questions, presentation, and topics are likely to be dramatically different for different audiences.
- Define the questions: Survey questions should be carefully defined to obtain the purposes of the survey. Careful focus on the purpose helps to keep the survey short, which increases the response rate and the quality of the answers. Be sure that questions are not leading, and that response options are balanced.
- Test the questions: Before administering a large survey, test the questions on a smaller sample of the target audience. If possible, administer the test questions in person, so that you can learn about problems with the questions by personal comments, rather than just the answers to the questions. Refine the questions until the test responders understand the questions the same way you do.
- Administer the survey: The survey can be delivered by a variety of methods, including in-person, mail, telephone, and the web.
- Follow up with non-responders: For mail, telephone, or web surveys you may find that some people fail to respond. Follow up with those who do not respond to try to increase the response rate.
- Analyze the result: Statistical analysis of the results is helpful, including identifying the limits of confidence on the answers.

See also

Development Reference: Focus Groups (11.31).

Further Reading

An excellent overview of processes for developing surveys is found in Dillman DA, Smyth JD, Christian LM (2014) Internet, phone, mail, and mixed-mode surveys: the tailored design method, 4th ed. Wiley, New York.

Surveys provide a method for obtaining market information from a large sample of people.

11.66 Theory of Inventive Problem Solving (TRIZ)

Find creative solutions to seemingly insoluble conflicts

TRIZ (pronounced TREES) is a Russian acronym for Theory of Inventive Problem Solving. TRIZ was developed in 1946 by Genrich Altshuller after leading a study of over 1.5 million patents. Altshuller found that similar problems had been solved in different technical fields using only a few dozen inventive principles. These inventive principles, which are part of the TRIZ method, can be used to help teams identify non-traditional solutions to problems – especially problems involving seemingly difficult to resolve conflicts.

Altshuller found that there were only 1250 typical conflicts and that they could be overcome with 40 inventive principles. He also found that there were only 39 engineering parameters that engineers tried to improve.

How to use TRIZ

1. Phrase the problem to reveal the conflict.
2. Identify the Engineering Parameters involved in the conflict.
3. Identify Inventive Principles (there are only 40) that could resolve the conflict.
4. Create ideas based on those inventive principles.

Example

A product development team worked on a next generation impact driver design. These drivers (similar to a drill driver) are often used by people in the construction industry to hang drywall. The team learned that the market required comfort and little down time because of dead batteries. Using a requirements matrix, the team translated these requirements into mass of the impact driver (this has a large impact on the comfort of using the driver), and power capacity in the driver batteries.

The team soon learned that greater power capacity for the batteries leads to a more massive impact driver – a conflict that is seemingly difficult to resolve.

Following the method above, the team identifies engineering parameters number 1 (weight of moving object) and 21 (power) as being involved in the conflict. With these two parameters in mind, the team identified an inventive principle that could resolve the conflict – inventive principle 1 (segmentation). Examples[1] of segmentation include (i) divide an object into independent parts, (ii) make an object sectional (for ease of assembly or disassembly), and (iii) increase the degree of an object's segmentation.

The team chose to divide the impact driver into two parts – the batteries, and the impact driver mechanism. In the end, they developed a drill driver with a battery in a backpack. More weight carried in a location that moves less resulted in a more comfortable drill driver with greater power capacity in the batteries.

See also

Development Reference: Brainstorming (11.5).

Further Reading

Altshuller G, Shulyak L, Rodman S, Fedoseev U (2005) 40 Principles: TRIZ keys to technical innovation. Techni-Cal Innovation Center.

Terninko J, Zusman A, Zlotin B (1998) Systematic innovation: an introduction to TRIZ (theory of inventive problem solving). CRC Press LLC.

[1] Admittedly, the inventive principles can be difficult to understand based only on short description given in the table. A good expanded discussion of each principle is given in (Altshuller and Shulyak, 1996)

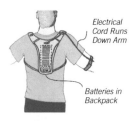

Electrical Cord Runs Down Arm

Batteries in Backpack

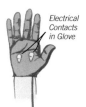

Electrical Contacts in Glove

Battery for handheld tool held in backpack. Tool powered by contacts in glove.

39 Engineering Parameters for TRIZ Method.

1. Weight of a moving object
2. Weight of a non-moving object
3. Length of a moving object
4. Length of a non-moving object
5. Area of a moving object
6. Area of a non-moving object
7. Volume of a moving object
8. Volume of a non-moving object
9. Speed
10. Force
11. Tension, pressure
12. Shape
13. Stability of object
14. Strength
15. Durability of a moving object
16. Durability of a non-moving object
17. Temperature
18. Brightness
19. Energy spent by a moving object
20. Energy spent by a non-moving object
21. Power
22. Waste of energy
23. Waste of substance
24. Loss of information
25. Waste of time
26. Amount of substance
27. Reliability
28. Accuracy of measurement
29. Accuracy of manufacturing
30. Harmful factors acting on object
31. Harmful side effects
32. Manufacturability
33. Convenience of use
34. Repairability
35. Adaptability
36. Complexity of device
37. Complexity of control
38. Level of automation
39. Productivity

40 Inventive Principles for TRIZ Method.

1. Segmentation
2. Extraction
3. Local quality
4. Asymmetry
5. Combining
6. Universality
7. Nesting
8. Counterweight
9. Prior counteraction
10. Prior action
11. Cushion in advance
12. Equipotentiality
13. Inversion
14. Spheroidality
15. Dynamicity
16. Partial or overdone action
17. Moving to a new direction
18. Mechanical vibration
19. Periodic action
20. Continuity of useful action
21. Rushing through
22. Convert harm into benefit
23. Feedback
24. Mediator
25. Self-service
26. Copying
27. Short-lived (disposable) instead of durable
28. Replacement of a mechanical system
29. Use of pneumatics or hydraulics
30. Flexible film or thin membranes
31. Use of porous material
32. Change the color
33. Homogeneity
34. Rejecting and regenerating parts
35. Transform physical and chemical states of an object
36. Phase transition
37. Thermal expansion
38. Use strong oxidizers
39. Inert environment
40. Composite materials

11.67 Troubleshooting

A systematic process helps fix non-working systems

Troubleshooting is the act of identifying and fixing problems that prevent an engineered system from working properly.

How to do it

The following steps form an effective troubleshooting process.

1. Prepare a troubleshooting log
   The troubleshooting log will contain a written record of all of the problems identified as well as all of the steps taken to solve the problem. It can be either hardcopy or electronic. However, it must be complete.
2. Document the problem
   A clear statement of the problem helps to both focus your efforts and know when the problem is finally fixed. An effective problem statement includes the following:

   a) What input are you providing to the system?
   b) What do you expect to happen?
   c) What actually happens?
   d) What steps are necessary to reproduce the problem?

   For intermittent problems, it is not uncommon that most of the time spent troubleshooting is spent identifying a reproducible method for causing the problem.
3. Obtain (or develop) a logical model of the system
   In order to know that something is wrong with your system, you need to know how the system is supposed to work. If you cannot obtain a logical model, you will need to create one. This often takes the form of a block diagram or a flowchart.
   The logical model should identify inputs and outputs for each of the elements in the model.
4. Narrow down the possible trouble location
   Following the procedures listed in step 2, cause the problem to occur.
   Divide the system in half. Check the first half of the system. Is it working properly? The logical model of the system should identify the desired outputs of the first half. If the first half isn't working properly, focus on the first half. Otherwise focus on the second half.
   Continue the subdivide and narrow process until you have identified a specific element that is causing the problem.
   Note that in some systems it may be necessary to physically separate the halves of the system when troubleshooting, as a failed component may affect signals throughout the system.
5. Repair or redesign the faulty element
   If the problem is due to a faulty physical component (e.g., a broken part), replace the component.
   If the problem is due to faulty engineering (e.g., a bug in a program), reengineer the component.
6. Verify that the change solved the problem
   Check to see that the new element is working properly. If this is the only fault in the system, the system will now work properly. Otherwise, return to step 4 and identify the next problem.

See also

Development Reference: Fault Tree Analysis (11.28).

Troubleshooting for even the most complex systems is facilitated by using a systematic troubleshooting process.

11.68 Uncertainty Analysis

Predict how changes in parameters change performance

Uncertainty analysis is a calculus-based method of determining how changes in one parameter will affect performance. It requires an analytical (closed-form) solution for the performance. When such a solution exists, uncertainty analysis is very quick and easy to do. If no such solution exists, *sensitivity analysis* is used to work with linear approximations of the solution.

Further Reading

NASA (2010) Measurement uncertainty analysis principles and methods: NASA measurement quality assurance handbook – annex 3. Available at http://www.hq.nasa.gov/office/codeq/doctree/NHBK873919-3.pdf

Bevington PR, Robinson DK (2002) Data reduction and error analysis for the physical sciences, 3rd ed. McGraw–Hill.

Wikipedia has an excellent article on uncertainty analysis: http://en.wikipedia.org/wiki/Experimental_uncertainty_analysis

Key methods

Uncertainty analysis uses partial derivatives to estimate the uncertainty in a given measurement. For a calculated quantity Q that depends on N measured values X_i, standard deviation of Q can be estimated from the standard deviation of the measurements:

$$\sigma_Q = \sqrt{\sum_{i=1}^{N} \frac{\partial Q}{\partial X_i} \sigma_{X_i}^2} \qquad (11.13)$$

This assumes that the individual measurements are independent, so that any measurement errors are uncorrelated. If the measurements are not independent, different formulas apply.

Uncertainty analysis can be used to understand the precision needed in manufacturing, if there is a functional relationship that governs the performance of the design. The standard deviation of the performance can be calculated from the design parameters in the same way the standard deviation of a calculated quantity can be calculated from the standard deviations of the measurements.

The process of determining the necessary precision in individual design parameters is known as tolerance allocation.

Applicability

Uncertainty analysis can always be used during validation. It is also used during subsystem design, and may be revisited when the design is refined during later stages of development.

See also

Development Reference: Design of Experiments (11.18), Robust Design (11.57), Sensitivity Analysis (11.61).

Uncertainty Analysis

11.69 Value Engineering

Focus on the parts of the design that add value to the customer

Value engineering is a design process focused on understanding the aspects of a product that provide value to a customer, and driving the design to create the best value. Value is provided to a customer by delivering product functions. Value is defined as the ratio of the benefit provided by the function to the cost of providing the function. So it is important to look at both increasing benefits and reducing costs.

Key methods

In value engineering, a product is analyzed in terms of the functions it provides. Each function a product provides is described in two words – an active verb followed by a measurable noun. Transmit force would be a valid function for value engineering. Orient bracket would not, because "bracket" is not a measurable noun, it's a component. For value engineering to reach its potential, we want to avoid describing the components of the product, as they are subject to change. Instead, we want to describe the functions, which are unchanging.

Value engineering recognizes two major categories of functions: basic functions and secondary functions. Basic functions are the functions that are necessary to perform the task that all products of this type must perform. For example, all fuel tanks must perform the function "contain fuel," so it is a basic function for a fuel tank.

Secondary functions are functions that are designed into a product to allow or enable the basic function to occur. They are further subdivided into the following categories:

Dependent Critical Functions: Functions that must occur in order to have the basic function occur.

Independent or Supporting Functions: Functions that help the basic function to be delivered better, faster, longer, etc. Virtually all of the competitive advantage for a particular product will be found in the supporting functions, because every product performs the basic functions.

All-the-time Functions: Functions that are requirements for the product but that are not generally related to the basic function, such as providing reliability, corrosion resistance, working in typical conditions, etc.

Design Criteria: Functions that are related to the important market requirements of the product that are not otherwise captured.

A key diagram to understand the functions provided by a product is the function analysis system technique (FAST) diagram. FAST diagrams capture the basic function, the dependent critical functions, the supporting functions, the design criteria, and the all-the-time functions in one brief diagram.

Applicability

Value Engineering is often applied more in redesign of existing products than in design of totally new products.

The FAST diagram is typically created during the opportunity development or concept development stages of development.

Once the FAST diagram is created, concept creation is carried out on key functions that are identified as opportunities to increase value.

Further Reading

Borza J, (2011) Fast diagrams: the foundation for creating effective function models. Proceedings of TRIZCON 2011 Detroit, Mi. Available for download from http://www.aitriz.org/documents/TRIZCON/Proceedings/2011-06_FAST-Diagrams-The-Foundation-for-Creating-Effective-Function-pdf

A variety of summary resources is available as a pdf from SAVE International, a professional society focused on value engineering. http://www.value-eng.org/education_publications_function_monographs.php

Value Engineering

See also

Development Reference: Quality Function Deployment (11.51), Six Sigma (11.62).

CD

SSE

PRR

Value Engineering

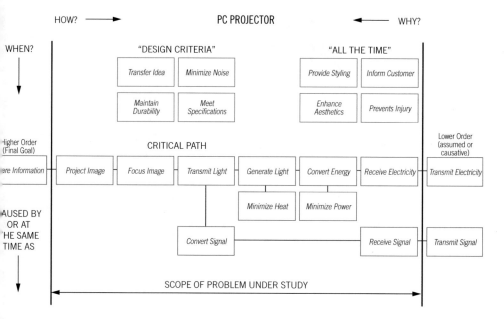

FAST diagram for a video projector, focusing on the basic function of the projector – sharing information. After orza.

APPENDIX **A**

Summary of Evolution in Product Development

This appendix contains a brief summary of the evolution that takes place during product development.

The six stages of product development are shown in Figure A.1. They are:

1. Opportunity development

2. Concept development

3. Subsystem engineering

4. System refinement

5. Producibility refinement

6. Post-release refinement

The stages of product development are fundamentally stages of design evolution, as the design moves from less detail to more detail. Along with the design, requirements and tests evolve in concert throughout the stages of development, as shown in Figure A.1. The evolution is tracked by observing changes in design artifacts that capture the team intent in a transferable manner.

Models and prototypes are testable representations of the design that are used

to test how well the design meets the requirements at a given state of evolution.

Figure A.2 shows how the requirements, tests, and design evolve in a coordinated, iterative fashion.

At the end of each stage of development, the design is submitted for review and approval. Major work on the subsequent stage is undertaken only after the results of the current stage have been approved.

Figure A.3 shows how desirability and transferability are evaluated during a development stage. It shows the relationships between artifact checking, performance testing, and validation testing.

Table A.1 summarizes the nature of information checks, performance tests, and validation tests. Before approval at each stage, the design must be demonstrated to have sufficient desirability and transferability. The demonstration must satisfy the project approvers, who are external to the development team.

The six stages of product development are summarized in Tables A.2 through A.7. The top-level activity maps for each stage are repeated in Figures A.4 through A.10

© Springer Nature Switzerland AG 2020
C. A. Mattson, C. D. Sorensen, *Product Development*,
https://doi.org/10.1007/978-3-030-14899-7

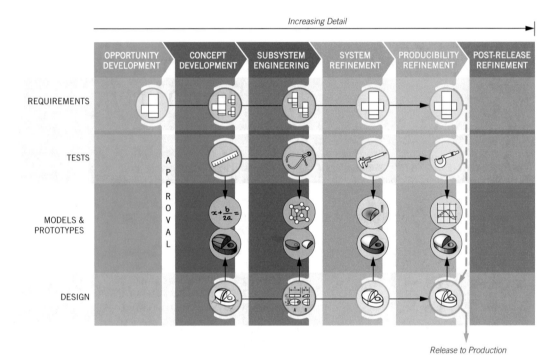

Figure A.1: The product requirements, tests, and design evolve through the stages of product development; the design is eventually used to manufacture the product. Prototypes and models are testable representations of the design used to determine the performance of the design at the current stage of evolution.

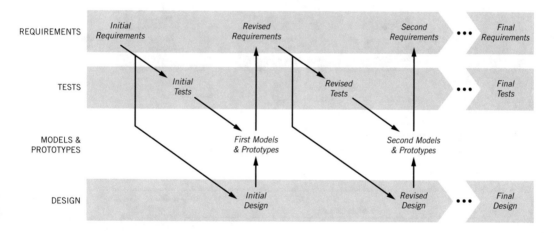

Figure A.2: The co-evolution of the requirements, the tests, and the design as aided by prototypes and models. Prototypes and models are created as a snapshot consistent with the current design and tested to measure and predict the performance of the design to compare with the requirements.

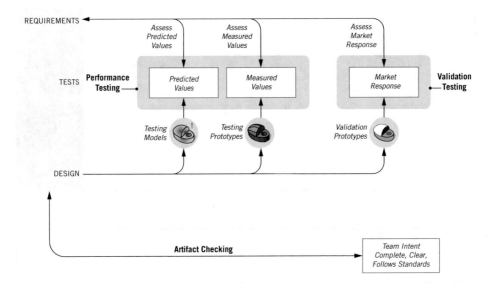

Figure A.3: Evaluation of desirability and transferability testing during a development stage. This shows the relationship between artifact checks, performance tests, and validation tests. The design is checked for quality and completeness. Using prototypes and models based on the design, product performance is measured or predicted using the tests. Validation testing consists of having a market representative to evaluate a prototype and pass judgment on the quality of the design. For approval, all required information should be present, the predicted and/or measured performance should meet the ideal or target values, and the market response should demonstrate that the market requirements are met.

Table A.1: Overview of desirability and transferability testing. This table summarizes the artifact checks, performance tests, and validation tests that must be successfully completed to obtain approval at the end of a product development stage.

	Artifact Checks	Performance Tests	Validation Tests
Test objective:	Check the quality and completeness of the design, requirements, and tests. Ensure that these fully and unambiguously capture what the team intended.	Use models and prototypes based on the design to predict and measure the performance of the design according to the tests. Compare the predicted and measured performance with the requirements.	Have the market representatives evaluate design artifacts to evaluate the quality of the design from the market perspective.
Artifacts used in testing:	Transferable medium appropriate for communicating the design (e.g., written list, technical drawing, bill of materials, test report, requirements matrix)	Models or prototypes based on the design; standard test methods that govern testing	Prototypes that communicate appropriately to the market representatives; standard test methods that govern validation tests
Artifacts created during testing:	Formal approval of the artifacts checked; usually includes release of a given revision	Test reports indicating the results of applying tests to models and prototypes; predicted and/or measured values (in a requirements matrix, for example)	Test reports indicating the results of validation tests; market response values (in a requirements matrix, for example)
Who is involved:	Evaluation performed by member(s) of the product development team, excluding the person who created the artifact	Evaluation performed by product development team	Evaluation coordinated by development team and performed by market representatives

Table A.2: Summary of the opportunity development stage.

Opportunity Development: Develop clear statements of market and engineering requirements that capture the market's desires for the product.

	Required information	Typical artifacts	Checking criteria	Approval criteria
Requirements	Market requirements	Section A of the requirements matrix	Consistent level of generality? Capture the most important requirements? Appropriate number of requirements? Reasonable differences in importance?	Complete and appropriate as evaluated by market representative
Requirements	Performance measures	Section B of the requirements matrix	Clearly measurable (even for subjective)? Capture market requirements well? Generally dependent, rather than independent? Units given and appropriate? Number appropriate? Appropriate importance ratings?	Market representatives (for less technical measures) and/or project approvers (for highly technical measures) find the measures to be appropriate.
Requirements	Requirement-measure correlations	Section C of the requirements matrix	All requirements have at least one measure? All measures have at least one requirement? More than just one-to-one correlations? Appropriate, defensible correlations?	Judged appropriate by project approvers
Requirements	Ideal values	Section D of the requirements matrix	Values make sense? Values are consistent with market requirements?	Judged appropriate by project approvers and market representatives
Tests	None required	Reports of team interactions with market representatives (e.g., surveys, interviews, etc.)	Are the interactions accurately conveyed in the report?	Not approved directly, but used to support approval of the requirements
Models	None required	Simple models that relate desirability to measured performance of competitors (e.g., screen size, battery life)	Do the models make logical sense?	Not approved directly, but used to support approval of the requirements
Prototypes	None required	Rough prototypes (foamboard, paper, foam, clay, cardboard, plywood, etc.) used to communicate with market representatives. Don't fully reflect eventual product design.	Do the prototypes facilitate communication with market representatives?	Not approved directly, but used to support approval of the requirements
Design	None required	Rough sketches or drawings of competitors or generic product possibilities. Don't fully reflect eventual product design.	Do the sketches or drawings facilitate communication with market representatives?	Not approved directly, but used to support approval of the requirements
	Useful tools:	Basic design process, competitive benchmarking, financial analysis, focus groups, interviews, observational studies, patent searches, planning canvas, project objective statement, quality function deployment, requirements matrix, surveys.		
	Common pitfalls:	Assuming, not validating; using only subjective performance measures; delaying feedback; devaluing the opportunity development stage; spending too much time.		

315

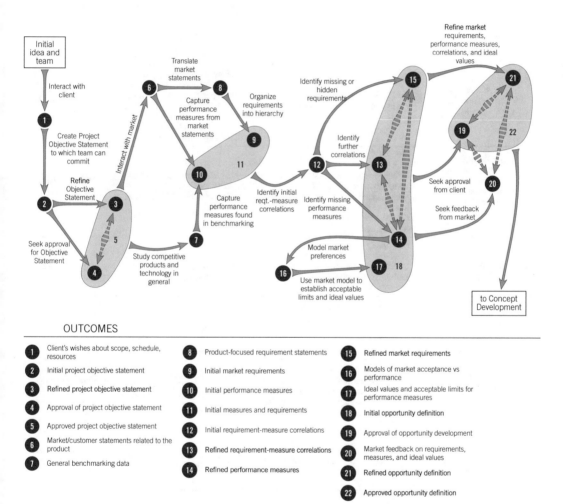

Figure A.4: Top-level activity map for the opportunity development stage.

Table A.3: Summary of the concept development stage.

Concept Development: Create a concept for the product and evolve it to have enough information to create basic estimates of cost, size, weight, and feasibility. Also include subsystem definitions, interface definitions, and target values for performance.

	Required information	Typical artifacts	Checking criteria	Approval criteria
Requirements	Subsystem requirements	Section A-D of the requirements matrix for each subsystem	See the Opportunity Development summary for the checking criteria for requirements matrices.	Complete and appropriate.
	Target values for the system and the subsystems.	Section E of the requirements matrix for the system and all subsystems.	Consistent with ideal values? Achievable with the selected product concept? Target values for all performance measures? Distinction between constraints, success measures, stretch goals? Trade-offs to less than ideal performance justified?	Consistent with expectations of market
Tests	Justification for concept selection, including model analysis and/or prototype test data demonstrating concept feasibility.	List, chart, or summary of considered concepts. Evaluation summary of considered concepts demonstrating feasibility of selected concept. Model and/or prototype test reports demonstrating the validity of the selected concept.	Are the test methods complete and correct? Is there enough detail for a third party to repeat the tests?	The desirability of the concept is demonstrated.
Models	Rough technical models of the product concept.	Low-fidelity fundamental models of the concept's operating principles. Statistical models describing the performance of experimental prototypes or related existing products.	Are the models reported in enough detail to allow a third party to use them?	Not directly approved; used with tests.
Prototypes	Simple prototypes of the product concept.	Rough prototypes (foamboard, paper, foam, clay, cardboard, plywood, etc.) used to show how the concept functions.	Are the prototypes appropriate for the intended use?	Not directly approved; used with tests.
Design	Geometric and other appropriate definition of the concept.	One or more of the following: Annotated hand sketches of concept. Overall system CAD model. Layout drawing. Skeleton drawing. Notes on sketches/drawings explaining concept. Block diagrams of electrical or fluid systems.	Is the design clearly communicated? Can a third party understand the intended design?	The design is sufficiently transferable to support cost, size, weight, and feasibility estimates.
	Decomposition of product concept into subsystems.	Tree or other relationship diagram that shows structure of decomposition. List of subsystems.	Are the subsystems and their relationships clearly shown?	The decomposition is appropriate for the selected concept.
	Subsystem interface definitions.	Interface matrix showing where subsystems interact. Product-focused requirement statements for each of the interfaces. Performance measures for each of the interfaces. Subteam responsibility assignments for each interface.	Is the interface matrix complete? Are the interface definitions complete? Are the interface definitions appropriate to achieve the desired performance?	The interfaces are fully defined and appropriate for the concept.
	Preliminary Bill of Materials.	Spreadsheet or database table that lists all known components, even if they have only a part name at this time.	Is the BOM complete at the level of known detail?	The bill of materials is appropriate.
	Useful tools:	Bill of materials, bio-inspired design, brainstorming, competitive benchmarking, controlled convergence, decomposition, internet research, interviews, literature review, method 635, mind maps, prototyping, recombination table, requirements matrix, scoring matrix, screening matrix, sketching, theory of inventive problem solving (TRIZ).		
	Common pitfalls:	Concept fixation; premature concept critique; reinventing the wheel; vague interfaces; decision delay.		

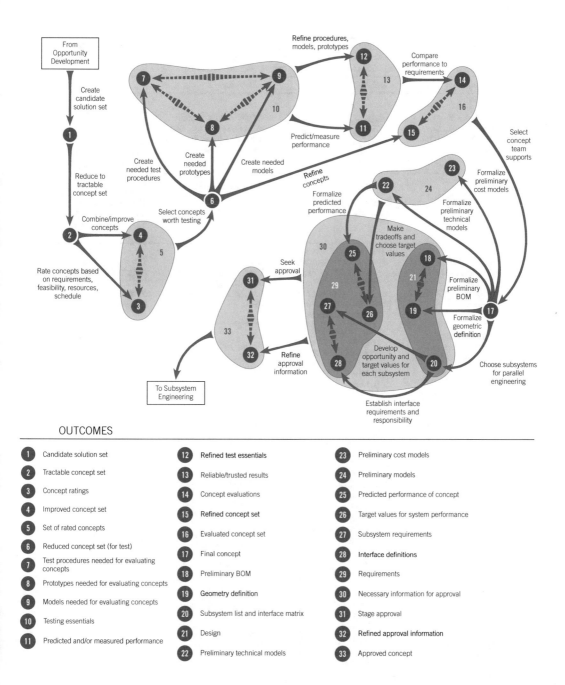

OUTCOMES

1	Candidate solution set	**12**	Refined test essentials	**23**	Preliminary cost models
2	Tractable concept set	**13**	Reliable/trusted results	**24**	Preliminary models
3	Concept ratings	**14**	Concept evaluations	**25**	Predicted performance of concept
4	Improved concept set	**15**	Refined concept set	**26**	Target values for system performance
5	Set of rated concepts	**16**	Evaluated concept set	**27**	Subsystem requirements
6	Reduced concept set (for test)	**17**	Final concept	**28**	Interface definitions
7	Test procedures needed for evaluating concepts	**18**	Preliminary BOM	**29**	Requirements
8	Prototypes needed for evaluating concepts	**19**	Geometry definition	**30**	Necessary information for approval
9	Models needed for evaluating concepts	**20**	Subsystem list and interface matrix	**31**	Stage approval
10	Testing essentials	**21**	Design	**32**	Refined approval information
11	Predicted and/or measured performance	**22**	Preliminary technical models	**33**	Approved concept

Figure A.5: Top-level activity map for the concept development stage.

Table A.4: Summary of the subsystem engineering stage.

Subsystem Engineering: Create high-quality engineered subsystems that have been demonstrated to be desirable. Design for other characteristics such as manufacturability and ergonomics should have been accomplished. At the end of this stage, the entire system has been designed, although the integration of subsystems has not yet been demonstrated.

	Required information	Typical artifacts	Checking criteria	Approval criteria
Requirements	Predicted and measured values for subsystem performance measures.	Section E of the requirements matrix for each subsystem	All predicted values are present, even if the value is N/A? All measured values present?	The subsystems meet or exceed the target values of the subsystem performance measures. If a few of the targets are not met, performance is at least in the acceptable range.
	Predicted values for system performance measures	Section E of the system requirements matrix	Predicted values for all performance measures? Consistent with subsystem measured values?	The predicted values meet or exceed the target values for the system performance measures. If a few of the targets are not met, performance is at least in the acceptable range.
Tests	Test data demonstrating subsystem desirability (measured values for subsystem performance measures).	Reports on methods and results demonstrating the desirability of the subsystem designs. Plots showing variation of subsystem performance with changes in design parameters.	Are the test methods complete and correct? Is there enough detail for a third party to repeat the tests?	The tests have been carried out with sufficient quality and transferability to provide strong evidence for the measured and predicted values.
Models	Engineering models used to choose design parameters and predict values for subsystem performance measures.	Software source code with run results. Input files for commercial software with run results. Excel spreadsheets. MathCAD worksheets. Hand solutions.	Are the models reported in enough detail to allow a third party to use them?	Not approved directly; used with tests.
Prototypes	Prototypes used to help choose values of design parameters. Prototypes used to measure measured values of subsystem performance measures.	Experimental testbeds used to select design parameter values. Fully functional subsystems for testing measured values.	Are the prototypes appropriate for the intended use? Is the fidelity and workmanship appropriate?	Not approved directly; used with tests.
Design	Geometric, material, and other appropriate definition of the design for each subsystem.	Engineering drawings of all custom-designed parts. Specifications (and possibly ordering information) for all purchased parts. Assembly drawings for the system and all subsystems. Schematic diagrams. Piping and/or wiring diagrams. Block diagrams. PC board layout files. Flowcharts and source code for any software that is part of the design.	Does the design package meet the design intent of the team? Are all relevant standards met? Are all the necessary components included? Is the design package sufficient to allow a third party to correctly make the product and test its compliance with specifications? Are all elements under version control?	The design is sufficiently transferable to allow the creation of complete subsystems and their integration into a complete system by a third party.
	Complete bill of materials	Spreadsheet or database table that lists all known components	Is it complete? Does it have all necessary information? Is it clear and unambiguous?	The bill of materials is appropriate
Useful tools:		Bill of materials, CAD modeling, checking drawings, design for assembly, design for manufacturing, design of experiments, design structure matrix, dimensional analysis, engineering drawings, ergonomics, experimentation, failure modes and effects analysis, fault tree analysis, finite element modeling, prototyping, sensitivity analysis, uncertainty analysis.		
Common pitfalls:		Avoiding analysis; reinventing analysis; never doing analysis; poor experimental procedure; focusing on the prototype, rather than the design.		

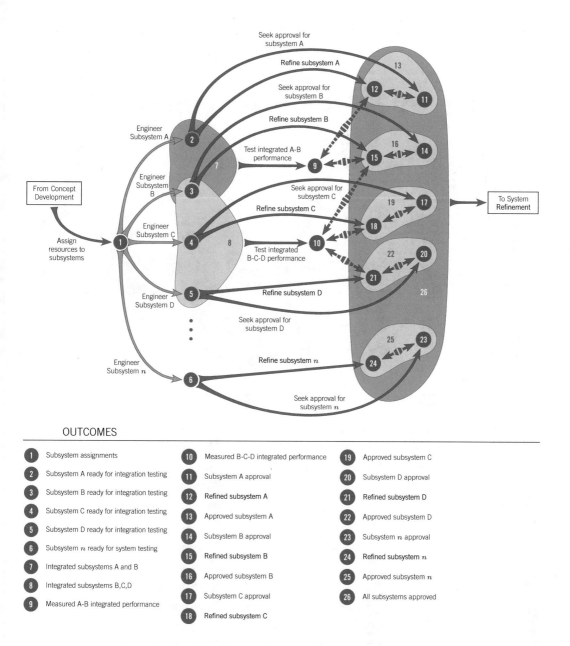

OUTCOMES

1 Subsystem assignments	**10** Measured B-C-D integrated performance	**19** Approved subsystem C
2 Subsystem A ready for integration testing	**11** Subsystem A approval	**20** Subsystem D approval
3 Subsystem B ready for integration testing	**12** Refined subsystem A	**21** Refined subsystem D
4 Subsystem C ready for integration testing	**13** Approved subsystem A	**22** Approved subsystem D
5 Subsystem D ready for integration testing	**14** Subsystem B approval	**23** Subsystem n approval
6 Subsystem n ready for system testing	**15** Refined subsystem B	**24** Refined subsystem n
7 Integrated subsystems A and B	**16** Approved subsystem B	**25** Approved subsystem n
8 Integrated subsystems B,C,D	**17** Subsystem C approval	**26** All subsystems approved
9 Measured A-B integrated performance	**18** Refined subsystem C	

Figure A.6: Top-level activity map for the subsystem engineering stage. There is a great deal of complexity not shown in this map related to the engineering of each individual subsystem. Please refer to Figure A.7 for the detailed top-level map for engineering a single subsystem.

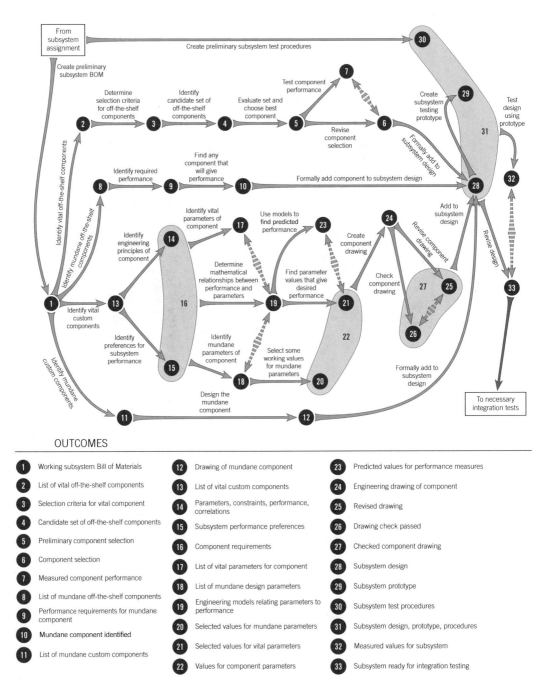

OUTCOMES

1	Working subsystem Bill of Materials	**12**	Drawing of mundane component	**23**	Predicted values for performance measures
2	List of vital off-the-shelf components	**13**	List of vital custom components	**24**	Engineering drawing of component
3	Selection criteria for vital component	**14**	Parameters, constraints, performance, correlations	**25**	Revised drawing
4	Candidate set of off-the-shelf components	**15**	Subsystem performance preferences	**26**	Drawing check passed
5	Preliminary component selection	**16**	Component requirements	**27**	Checked component drawing
6	Component selection	**17**	List of vital parameters for component	**28**	Subsystem design
7	Measured component performance	**18**	List of mundane design parameters	**29**	Subsystem prototype
8	List of mundane off-the-shelf components	**19**	Engineering models relating parameters to performance	**30**	Subsystem test procedures
9	Performance requirements for mundane component	**20**	Selected values for mundane parameters	**31**	Subsystem design, prototype, procedures
10	Mundane component identified	**21**	Selected values for vital parameters	**32**	Measured values for subsystem
11	List of mundane custom components	**22**	Values for component parameters	**33**	Subsystem ready for integration testing

Figure A.7: Top-level activity map for engineering a single subsystem. This map will be repeated for each subsystem. Included in this map are submaps dealing with mundane off-the-shelf components, vital off-the-shelf components, mundane custom components, and vital custom components. Note that a given subsystem may have zero, one, or more than one of any or all of these kinds of components.

Table A.5: Summary of the system refinement stage.

System Refinement: Integrate the subsystems into a demonstrated, high-quality working system. Refine the design as necessary to resolve any difficulties encountered during testing.

	Required information	Typical artifacts	Checking criteria	Approval criteria
Requirements	Measured system performance values	Section E of the system requirements matrix	Are all predicted values present, even if the value is N/A? Are all measured values present?	The system meets or exceeds the target values of the system performance measures. If a few of the targets are not met, performance is at least in the acceptable range.
	Market response to the product	Section F of the requirements matrix. Reports on customer response to the product, as measured by surveys, focus groups, interviews, or other direct interaction.	Has the market response been adequately assessed?	The market finds the product desirable.
Tests	Updated methods used to predict and measure the performance of the entire system (or product).	Reports on methods and results demonstrating the desirability of the product. Plots showing variation of system performance with changes in design parameters.	Are the test methods complete and correct? Is there enough detail for a third party to repeat the tests?	The tests have been carried out with sufficient quality and transferability to provide strong evidence for the measured and predicted values.
Models	Engineering models used to choose design parameters and predict values for system performance measures.	Model source code with run results. Input files for commercial software with run results. Excel spreadsheets. MathCAD worksheets. Hand solutions.	Are the models reported in enough detail to allow a third party to use them?	Not approved directly; used with tests.
Prototypes	Prototypes used to help choose values of design parameters. Prototypes used to determine measured values of system performance measures.	Experimental testbeds used to select design parameter values. Fully functional systems for testing measured values.	Are the prototypes appropriate for the intended use? Is the fidelity and workmanship appropriate?	Not approved directly; used with tests.
Design	Refined definition for the entire system.	Engineering drawings of all custom-designed parts. Specifications (and possibly ordering information) for all purchased parts. Subassembly and assembly drawings and instructions for all subsystems and the system. Schematic diagrams. Piping and/or wiring diagrams. Block diagrams. PC board layout files. Flowcharts and source code for any software that is part of the design.	Does the design package meet the design intent of the team? Are all relevant standards met? Are all the necessary components included? Is the design package sufficient to allow a third party to correctly make the product and test its compliance with specifications? Are all elements under revision control?	The design is sufficiently transferable to support the creation of the entire system by a third party.
	Complete bill of materials	Spreadsheet or database table that lists all known components	Is it complete? Does it have all necessary information? Is it clear and unambiguous?	The bill of materials is appropriate
Useful tools:		Bill of materials, CAD modeling, checking drawings, design for assembly, design for manufacturing, design of experiments, design structure matrix, dimensional analysis, engineering drawings, experimentation, failure modes and effects analysis, fault tree analysis, finite element modeling, prototyping, sensitivity analysis, uncertainty analysis.		
Common pitfalls:		Following poor experimental procedure; creating new and distracting performance measures; making poor trade-offs; creating new problems; substituting team judgment for the market.		

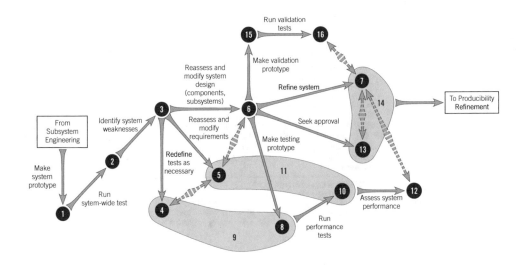

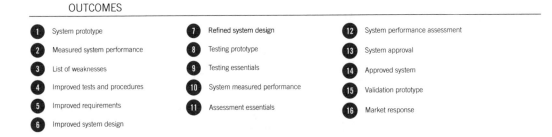

OUTCOMES

1 System prototype	**7** Refined system design	**12** System performance assessment
2 Measured system performance	**8** Testing prototype	**13** System approval
3 List of weaknesses	**9** Testing essentials	**14** Approved system
4 Improved tests and procedures	**10** System measured performance	**15** Validation prototype
5 Improved requirements	**11** Assessment essentials	**16** Market response
6 Improved system design		

Figure A.8: Top-level activity map for system refinement.

Table A.6: Summary of the producibility refinement stage.

	Required information	Typical artifacts	Checking criteria	Approval criteria
Producibility Refinement: Refine the design as necessary to allow a desirable product to be produced in the desired quality and quantity. Note that this stage is primarily about fixing producibility weaknesses that are identified during production ramp-up.				
Requirements	Updated predicted and measured values of performance measures.	Part E of the system and subsystem requirements matrix.	Are all predicted values present, even if the value is N/A? Are all measured values present?	The system meets or exceeds the target values of the system performance measures. If a few of the targets are not met, performance is at least in the acceptable range.
Tests	Updated methods used to predict and measure the performance of the system. Methods used to demonstrate the producibility of the product.	Reports on methods and results demonstrating the desirability of the product. Reports of producibility challenges and their resolution.	Are the test methods complete and correct? Is there enough detail for a third party to repeat the tests?	The tests have been carried out with sufficient quality and transferability to provide strong evidence for the measured and predicted values.
Models	Engineering models used to analyze and adjust design characteristics related to producibility. Statistical analysis of producibility challenges (defects, low rate, high cost, etc.).	Model source code with run results. Input files for commercial software with run results. Excel spreadsheets. MathCAD worksheets. Hand solutions.	Are the models reported in enough detail to allow a third party to use them?	Not approved directly; used with tests.
Prototypes	Prototypes used to explore possible producibility solutions. Prototypes used to measure producibility of revised design.	Pilot-scale production runs with statistical analysis. Producibility studies on initial product runs and refined design.	Do the prototypes demonstrate the solutions to producibility problems?	Not approved directly; used with tests.
Design	Refined definition for the entire system.	Engineering drawings of all custom-designed parts. Specifications (and possibly ordering information) for all purchased parts. Subassembly and assembly drawings and instructions for all subsystems and the system. Schematic diagrams. Piping and/or wiring diagrams. Block diagrams. PC board layout files. Flowcharts and source code for any software that is part of the design.	Does the design package meet the design intent of the team? Are all relevant standards met? Are all the necessary components included? Is the design package sufficient to allow a third party to correctly make the product and test its compliance with specifications? Are all elements under revision control?	The design is sufficiently transferable to support production by a third party.
	Complete bill of materials	Spreadsheet or database table that lists all known components	Is it complete? Does it have all necessary information? Is it clear and unambiguous?	The bill of materials is appropriate
Useful tools:	Design for assembly, design for manufacturing, design of experiments, experimentation, failure modes and effects analysis, fault tree analysis, plan-do-check-act, sensitivity analysis, six sigma, theory of inventive problem solving, troubleshooting, uncertainty analysis.			
Common pitfalls:	Creating new problems; inadequate sample sizes; waiting until this stage to consider producibility.			

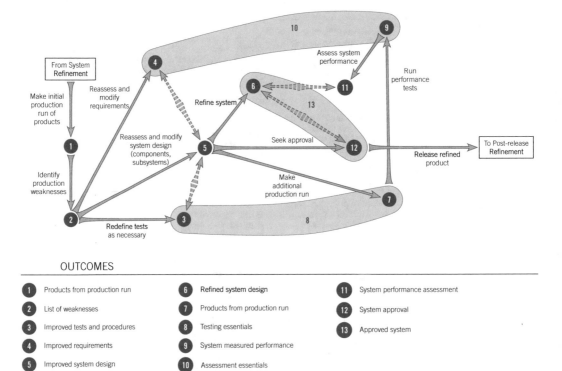

OUTCOMES

1	Products from production run	**6**	Refined system design	**11**	System performance assessment
2	List of weaknesses	**7**	Products from production run	**12**	System approval
3	Improved tests and procedures	**8**	Testing essentials	**13**	Approved system
4	Improved requirements	**9**	System measured performance		
5	Improved system design	**10**	Assessment essentials		

Figure A.9: Top-level activity map for producibility refinement.

Table A.7: Summary of the post-release refinement stage.

Post-release Refinement: Refine the design to improve the desirability of the mass-produced product. This includes items such as decreasing cost, increasing functionality, and eliminating weaknesses that become apparent after the product has been offered to the market.

	Required information	Typical artifacts	Checking criteria	Approval criteria
Requirements	Updated market and product requirements, including the additional information that was learned during this stage.	Updated subsystem and system requirements matrices.	Are all of the matrices complete and correct?	The changes in the market and product requirements capture the market's desires. The system meets or exceeds the target values of the system performance measures. If a few of the targets are not met, performance is at least in the acceptable range.
Tests	Updated methods used to predict and measure the performance of the system.	Reports on methods and results demonstrating the desirability of the product.	Are the test methods complete and correct? Is there enough detail for a third party to repeat the tests?	The tests have been carried out with sufficient quality and transferability to provide strong evidence for the measured and predicted values.
Models	Engineering models used to analyze and adjust design characteristics of the product. Statistical analysis of product weaknesses (defects, low rate, high cost, etc.)	Model source code with run results. Input files for commercial software with run results. Excel spreadsheets. MathCAD worksheets. Hand solutions.	Are the models reported in enough detail to allow a third party to use them?	Not approved directly; used with tests.
Prototypes	Prototypes used to explore possible improvements. Prototypes used to measure producibility of revised design.	Production runs with statistical analysis.	Do the prototypes demonstrate both the problem and the solutions?	Not approved directly; used with tests.
Design	Refined definition for the entire system.	Engineering drawings of all custom-designed parts. Specifications (and possibly ordering information) for all purchased parts. Subassembly and assembly drawings and instructions for all subsystems and the system. Schematic diagrams. Piping and/or wiring diagrams. Block diagrams. PC board layout files. Flowcharts and source code for any software that is part of the design.	Does the design package meet the design intent of the team? Are all relevant standards met? Are all the necessary components included? Is the design package sufficient to allow a third party to correctly make the product and test its compliance with specifications? Are all elements under version control?	The design is sufficiently transferable to support production by a third party.
	Complete bill of materials	Spreadsheet or database table that lists all known components	Is it complete? Does it have all necessary information? Is it clear and unambiguous?	The bill of materials is appropriate.
	Useful tools:	Design for assembly, design for manufacturing, design of experiments, experimentation, failure modes and effects analysis, fault tree analysis, plan-do-check-act, sensitivity analysis, six sigma, theory of inventive problem solving, troubleshooting, uncertainty analysis value engineering.		
	Common pitfalls:	Ignoring the market; creating new problems; inadequate sample sizes; failure to fully document everything.		

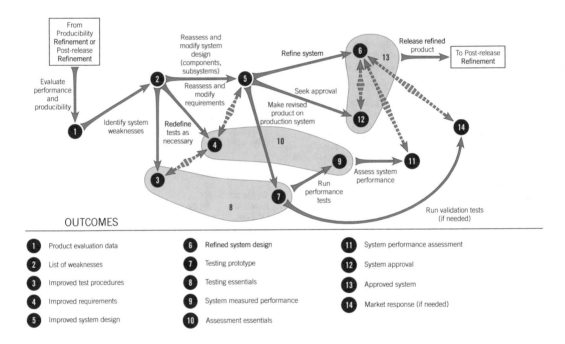

Figure A.10: Top-level activity map for post-release refinement.

B

Human-Powered Water Well Drill

The following case study describes a product development effort to bring clean drinking water to rural communities in sub-Saharan Africa. A more detailed case study on the village drill that articulates the introduction of the village drill into the market can be found at C.A. Mattson, A.E. Wood, and J. Renouard, "Village Drill: A Case Study in Engineering for Global Development With Five Years of Data Post Market-Introduction," 2017, Journal of Mechanical Design, Volume 139, Issue 6, pages 065001 (10 pages), doi: 10.1115/1.4036304. This development effort resulted in a human-powered machine that can drill a 6 inch diameter hole 250 feet into the ground to access clean drinking water. Throughout this book, the machine will be referred to as the *human-powered water well drill*, or the *drill* for short.

The purpose of this case study is to provide enough details regarding the drill so that the product development theory presented in Part I of this book can be discussed in the context of a real product (the drill). We choose to use the drill as a case study because we have access to nearly all the information related to its development. We note that in some cases we have made minor alterations to the development team's original work in order to maintain consistency with the language used throughout other parts of this book[1].

Project Client: WHOlives.org

Engineering Development Team: WaterDOT, a Brigham Young University Capstone team.

Project Dates: September 2010 – May 2011.

Project Scope: Middle of the opportunity development stage to the end of the system refinement stage.

Eight-Year Update: Product is in the post-release refinement stage. Eighty-seven drills have been produced, and are being used in 33 countries across Africa, Asia, and Latin America. More than 3,400 wells have been drilled. More than 2.5 million people are getting water from those wells.

B.1 Case Study Introduction

Tanzania is one of the many countries in the world that suffers from extreme poverty.

---

[1] Much of the information in this chapter was produced by the development team as a final project report. To that end, we recognize the team members as Devin LeBaron, Eric Janmohamed, Jimmy Stacey, Ken Langely, Nathan Toone, Sabin Gautam, and Christopher Mattson (coach).

Figure B.1: The full system design being tested in Tanzania.

Many of the hardships in Tanzania can be attributed to the lack of clean water. Despite the fact that the country is surrounded by three major lakes and an ocean, and 7% of its area is covered by fresh water, it is difficult to find clean water because the water is contaminated and not suitable for human consumption.

Potable, or drinkable, water is the basis for a better life. It is estimated that Tanzanian women and children spend an average of 2 hours a day just collecting water, and it is common to find people who walk 6 hours just to find water. In addition to the time concerns, 80% of all disease in developing countries is caused by bad water. Many of these people die because of the lack of medicine and health care. Since these people are collecting contaminated water, they spend their time being sick, visiting doctors, and paying for medicine they cannot afford. Although the people know the water makes them sick, they have no alternative. Installing a village water well dramatically reduces all of these concerns

and provides clean water for more than 750 people per well.

Unfortunately, many villages lack clean water wells because the current methods of drilling in Tanzania are limited to opposite extremes. On one extreme, the drilling is done by a professional drilling rig, which is too expensive (from 7,000 to 15,000 USD), while the other extreme is a homemade drilling system, which is unsuccessful drilling beyond 150 feet, where potable water is often reached.

Of course, a professional drilling rig can drill to depths sufficient enough to access clean drinking water, but it costs upwards of ten thousand dollars to hire the rig for the few days required to drill the borehole. The villages that need these wells cannot afford to spend this extreme amount of money. As a result, they turn to homemade drilling systems, which often are insufficient. The primitive, manual methods with which they drill simply cannot drill deep enough to access clean water. The two main homemade (both manual) methods in Tanzania are Hand Augering and Rota-sludge. Hand Augering uses an auger to dig the earth away and is effective only in soft soil formations, reaching depths of no more than 125 ft. Rota-sludge is a more effective method because it reaches the same depths, but is able to work in more diverse soil formations. However, due to limited mechanical advantage and strength of tools, both of these methods are more practically limited to depths of less than 100 feet.

In mid-2010, a non-profit organization called WHOlives.org[2] was organized for the purpose of providing greater access to clean water, better health, and to provide more opportunity to those who lack the financial means of achieving it on their own. Accordingly, they have commissioned team WaterDOT to develop a way to bridge the gap between the village and the water by providing a manual water well drill that will enable the people to drill their own wells for roughly US$1,500. Intended mainly for developing countries such as Tanzania, the design needs to be affordable, and also extremely simple, as no product support or spare parts will be available.

B.2   Project Summary

Team WaterDOT has developed a human-powered water well drill that will provide access to clean drinking water to villages in Tanzania at an affordable cost. A prototype that represents the design at the end of the system refinement stage is shown in Figure B.1.

During the opportunity development stage the team discovered that the project would need to be developed in less than 1700 man hours and cost less than US$2,800 to manufacture all prototypes combined. They also discovered that in order to maximize the likelihood of reaching potable water, the six-inch borehole would need to reach a depth of 250 ft and cut through

various soil formations. In an effort to bridge the gap between expensive professional rigs and less effective homemade systems, it was established that the drill would use existing drill pipe and bits from the oil-drilling industry, operate primarily on manual power, and be portable to move from village to village. Further, the team determined that in order to break through the rock formations, the drill would need to supply up to 2,000 pounds of downward force, mostly from the drill pipe itself, as well as apply a maximum of 1000 foot-pounds of torque to the drill pipe.

Team WaterDOT successfully created a design that meets the above requirements. By the end of the project, the design consisted of three major components: the structure, the wheel support, and the wheel. The structure was designed to withstand loads of up to 6750 pounds before yielding, which is over three times the weight of 250 feet of drill pipe. The structure was designed with a low center of mass to prevent tipping and add stability to the drilling process. The lifting of the pipe is accomplished through the use of a winch and pulley system. This also allows the operators to control the penetration rate of the drill bit. The wheel support is able to stabilize and support the weight of the wheel (200 pounds). The wheel support also allows ready access to the borehole and the drill pipe underneath the wheel. The innovative design of the wheel consists of a hub that is permanently attached to the wheel support and 8 removable spokes. Each of the spokes is pinned in place on the hub, and additional strength is gained from two-foot-long cross braces that are placed between the spokes. This design allows for easy transportation and allows workers to remove spokes for improved access to the drill string.

In addition to meeting the quantitative specifications for drilling a borehole, the team's final design also meets the economic specifications. It is able to be manufactured for approximately US$1,600, and since the design consists mostly of welded steel, the majority of manufacturing

---

[2]The WHO in WHOlives.org stands for Water, Health, and Opportunity.

can be performed in Tanzania. The entire drilling rig can also be easily disassembled and packed into the bed of a regular sized truck or small trailer for transportation to remote areas of Tanzania.

WaterDOT has verified this design in both theory and reality. Throughout the design process the team has conducted many tests to determine the acceptability and the feasibility of each concept. The testing in the USA culminated in a final test with the fully functional steel prototype in which a 27 foot hole[3] was drilled in a sandy soil condition. Including setup, drilling, and cleanup, the entire test was completed in a 5 hour period. The system refinement tests in Tanzania involved drilling over 26 feet through significant rock formations and 130 ft in sandy soil conditions. These system refinement tests took 7 days of drilling.

In all, team WaterDOT has successfully designed a human-powered water well drill that can be implemented in Tanzania. This drill is substantially more robust than other manual drilling methods in use in Tanzania. Also, this drill can be manufactured and operated for a fraction of the cost of drilling one well with a professional drilling rig.

B.3   Main Requirements

In September 2010 the founders of WHOlives.org described their desires for the outcomes of this project. They explained their work in Tanzania and their plans for the future, including their vision of bringing water, health, and opportunity to the people. The team learned that WHOlives.org plans to produce the drills for a manufacturing cost lower than US$5,000, and lease the drills to villages for US$1,500 per successful borehole. Under the assumption that villagers are motivated to improve their village, WHOlives.org plans to have local villagers operate the drill without compensation. The revenue generated from the leases would pay for the drill manufacturing, the drill maintenance, a local trainer–leaser agent, and other operational costs of the organization.

The founders of WHOlives.org shared what they expect from the drill and its capabilities. From this meeting, the main requirements became evident. In addition to the needs outlined by the client, members of the development team discovered other important needs through research and visiting with well drilling experts. The requirements were evaluated by the team and written in a form that facilitated the creation of the requirements matrix. Figure B.2 shows a reduced[4] version of the matrix. The reduced matrix shows what the team believed to be the most important requirements.

The team was able to successfully meet the main requirements outlined at the beginning of the project. As a result, the team's final design was ready to proceed through the producibility refinement stage by WHOlives.org and their manufacturing partner in Africa. During the earlier stages of product development, the team put considerable effort into developing a design that could be easily manufactured in Tanzania. The development team believes that since the final design consists of mostly steel parts that are welded together, the drilling rig can be manufactured in Tanzania without the need of expensive and specialized machining equipment (excluding the pulleys, bearing, and winch, which are purchased parts).

The entire drilling rig can also be easily disassembled and transported in the bed of a regular size pick-up truck, which is approximately 5.5 feet wide and 7 feet long. This falls well within the ideal value of being transported on a 6 foot by 10 foot trailer. As a whole, the final design costs approximately $1,600 to manufacture. This manufacturing cost fell well below the requirement that manufacturing costs be below $5,000.

B.4   The Team's Final Design

The final design is described by the team as three major subsystems: the structure,

---

[3]Without a well driller's permit, the team was not allowed to drill beyond 30 feet in the USA.

[4]The matrix has been reduced to maintain readability in the format of this book; the team's original matrix had 68 market requirements and 83 performance measures.

Product: DRILL
Subsystem: N/A

Performance Measures

#	Performance Measure	Units	Importance	Lower Acceptable	Ideal	Upper Acceptable
1	Maximum borehole depth	ft	9	100	250	–
2	Time required to cut through 6 inches of rock	min	10	–	45	60
3	Downward drilling force	lbs	10	500	3,000	–
4	Torque applied to drill bit	ft-lbs	10	200	400	–
5	Compatable with X% existing drill bits	%	3	90	100	–
6	Water pressure down the pipe	psi	6	50	113	5
7	Percentage of water that leaks through sides	%	3	–	0	5
8	Percentage volume of cuttings removed	%	3	95	100	–
9	Depth cut per 8 hours of drilling	ft	3	4	36	–
10	Number of required people	people	9	–	3	12
11	Weight of heaviest subassembly	lbs	9	–	50	400
12	Longest dimension (l,w,h) of biggest subassembly	in	9	–	48	96
13	Percentage of drill manufacturable in Tanzania	%	9	85	100	–
14	Cost to produce 1 drill after development	USD	9	–	1,000	5,000
15	Time required to learn how to operate	hr	9	–	4	20
16	Height of hand operated parts	ft	1	2	3.5	5
17	Feels comfortable	n/a	1	Drill can be operated with occasional rest, and requires awkward movements that leave the user sore.	Drill can be operated continuously without the need to rest. Does not require awkward movements.	–
18	The Drill is attractive	n/a	4	Drill looks like a piece of machinery for drilling holes.	Drill has a professional look. People are interested in looking at it.	–
19	The Drill interests investors	n/a	12	The drill, when explained to investors, is something they want to invest in.	Drill captures media attention. Investor's are proactive in contributing. Drill has iconic look.	–

Market Requirements (Whats) — Importance

#	Market Requirement	Importance	Related Performance Measures
1	The Drill reaches potable water beyond 100 ft	9	1, 2
2	The Drill cuts through rock	1	2, 3, 4
3	The Drill uses existing drill bits	3	5
4	The Drill seals borehole sides to prevent cave-in	3	6, 7
5	The Drill removes cuttings from the borehole	3	6, 8
6	The Drill works at an efficient speed	3	9
7	The Drill uses only manual labor to function	9	2, 3, 11
8	The Drill is affordable	9	13, 14, 19
9	The Drill requires simple training to operate	9	15
10	The Drill is portable	9	11, 12
11	The Drill is comfortable to operate	1	16, 17
12	The Drill is attractive	1	18
13	The Drill attracts investors	3	18, 19

Figure B.2: Requirements matrix (reduced) for drill.

the wheel support, and the wheel. The team's final design is shown in Figure B.3. The square, yellow pipe shown in the figure descending through the wheel is known as a kelly bar. Since the kelly bar is square, it allows the pipe to be gripped by the wheel and to descend into the borehole simultaneously. The use of this technology from rotary table drilling had a large influence on the form of the rest of the design. In the subsequent sections each component will be described in detail along with the engineering analysis that accompanied the design.

Figure B.3: CAD rendering of team's final design. This represents the design at the end of the system refinement stage of product development.

Structure

The structure is composed of 4 parts: a base, two vertical columns, and a cantilevered beam for lifting the pipe (see Figure B.3). As depicted in figure, the base has two horizontal legs sufficiently long and wide enough apart to keep the structure balanced and stable.

Overall, the base is 47 inches wide and 84 inches long. It is made of 3.5 inch square tubing, 3/8 inch wall thickness. The size and mass of the base structure keep the center of mass for the whole structure low to prevent tipping over. In order to tip, the structure has to rotate 36.7 degrees from the vertical. To cause this rotation a horizontal force of 220 pounds must be applied to the high end of the cantilever beam, or a force of 352 pounds must be applied at the top of the 5 foot column; the likelihood that these large forces will be applied to the structure is extremely low. The two columns are 3 inch square tubes, with 1/4 inch wall thickness. This allows enough clearance to slide into the sleeves, yet strong enough to withstand the applied loads.

The cantilevered beam is a 5 inch square steel tube that is 7 feet long with a wall thickness of 3/16 of an inch. The beam has two sleeves of 3.5 inch steel tubing welded at a 45 degree angle that allow the beam to be slid securely on top of the vertical columns. The beam is designed to be pinned to the columns by four 4 inch long clevis pins. The high end is 9 feet above the ground, directly above the borehole. Both ends of the beam have a pulley inside, and a winch is attached to the low end of the beam. The stranded wire cable from the winch goes through the beam and can then hook onto the pipe/kelly bar for lifting.

A series of rectangular steel tubing sections are welded between the legs of the base over the borehole for additional support. They also provide a rest for the slip plate, which is used to secure the pipe while adding or removing pipe sections (this is called *changeover*).

Structural Analysis of Lifting System

The requirement for the lifting system is to be able to support and lift the weight of 250 feet of drill pipe. Based on the density of steel (490.6 pounds per cubic foot), a pipe wall thickness of 0.25 inches, and an outer diameter of 2.875 inches, the weight of 250 feet of pipe is 1,725 pounds. While drilling, the borehole may cave in on top of the pipe, thus necessitating the ability to lift more than the just the weight of the drill pipe.

The three major components of the lifting system are the hoist structure, the winch, and the pulleys. Of these components the most critical component is the hoist structure. During development the team ensured that the hoist would not yield, even under extreme lifting conditions. Because of the length of the cantilevered beam, the highest stresses occur in the beam at the junction with the first column. This stress is due to a combined bending load and axial load. Therefore, to select the appropriate beam size, the von Mises stresses were calculated at this point. A simple optimization program was created in Excel to optimize the beam dimensions given a load, a safety factor, and a beam wall thickness. From this optimization routine a 5 inch square steel beam was chosen with the yield strength of steel as 50,000 psi, a

safety factor of 1.5, a wall thickness of 0.188 inches, and a vertical load of 4,500 pounds.

The winch was then chosen to be able to lift the weight of the pipe and more. The team wanted to ensure that there would never be any failure of the lifting structure. A hand winch with a 3,500 pound first layer capacity (and a 1,849 pound full drum capacity) was selected. This winch has an enclosed gear for protection from the harsh environments of drilling and an automatic brake, which means that it cannot move unless an operator is rotating the handle even with tension in the wire rope. Furthermore, at its maximum capacity the operator only has to apply 19.4 pounds of force to the end of the winch handle to move the load.

Pulleys were selected to match the lifting capabilities of the winch as close as possible; however, the pulleys were also constrained in size by the inside dimension of the beam. Stainless steel pulleys with a 4.25 inch diameter and plain bronze bushings were selected. These pulleys have an operating capacity of 3,000 pounds.

Wheel Support

The wheel support is significantly different than previous designs considered by the team. It is made of several 3 inch by 2 inch rectangular steel tubing sections that are welded together to make a platform on one end that the lazy susan bearing and wheel can rest on (see Figure B.3). The other end has sections of tubing spaced wide enough to fit over the vertical columns of the structure. Two parallel sections slide around both columns and are bolted in place. Then two smaller sections six inches above the long sections slide around the long column only and are bolted in place. Bolting the wheel support to the columns in this manner provides more structural stability to the structure and the wheel support.

The platform end is 44.75 inches from the ground, which will make it ergonomically ideal for an average height operator to turn the wheel. The platform is 12 inches wide with ample space in the middle for the kelly bar and pipe to slide through. The only load that will be placed on the wheel support is the weight of the wheel itself.

Wheel

The wheel is made up of a central hub and eight spokes (see Figure B.3). The hub consists of eight 4 inch long sections of 3 inch by 2 inch rectangular steel tubing that are spaced evenly in an octagonal pattern with the open ends facing outward. These form sleeves for the spokes to be inserted and are sandwiched between two 1/4 inch thick octagonal plates that are 12 inches wide. The plates have 4.1 inch square holes in the middle that are aligned for the kelly bar to slide through. All components of the hub are welded together for robustness. A small piece of metal is welded to the inside bottom lip of each of the open tubing sections of the hub to prevent the spokes from sagging. The wheel hub is then attached to the wheel support by a thrust bearing allowing the wheel to spin freely.

The spokes are 1.5 inch by 2.5 inch rectangular tubing sections with a length of 3 feet. One end fits into one of the sleeves of the hub and is pinned in place. The other end has a handle consisting of an 11.5 inch long by 1.25 inch diameter solid steel rod going through the middle perpendicular to the main axis of the spoke. The diameter of 1.25 inches is ergonomically optimal for a power grip. The handle is centered on the spoke with 5 inches protruding on each side of the spoke. The purpose of this is to accommodate people of different heights working on the drill. The outside end of the spoke is closed and deburred for safety. For additional support of the wheel spokes, a 2 foot piece of 1 inch by 1 inch by 1/8 inch angle iron is pinned as a cross brace between all the spokes.

The six foot diameter of the wheel provides enough torque to drill efficiently in all soil types while still maintaining its portability. The spokes are not permanently attached to the hub so that the wheel may easily be

assembled and taken apart for transportation. Additionally, the weight of the wheel, especially the solid steel handles at the end of the spokes, provides enough inertia for the wheel to maintain a continuous motion and act as a flywheel.

Twist and Torsion of Drill Pipe

With the wheel applying a constant torque to the drill pipe, it is possible that some angle of twist will develop through the length of the drill pipe. This can cause unwanted wind-up that could potentially be dangerous if the wheel were suddenly released. Therefore, calculations were performed to determine the twist angle with 250 feet of pipe and a maximum torque of 1000 foot-pounds, which corresponds to three operators exerting 111 pounds of force at the edge of the wheel. In the limiting case where the drill bit is held stationary, 49 degrees of twist will develop in the pipe. This would result in the wheel unwinding a bit more than 1/8 of a turn, which means that at most one spoke will pass by the operator. In addition, with use of a winch and the subsequent upward force that can be applied to the pipe, the situation in which the drill bit is held stationary can be avoided.

Change-over Process

The change-over process has gone through many improvements since the team's previous designs. The largest change that has occurred is changing the 6 foot sections of pipe to 3 foot sections, and the kelly bar has also been reduced from 7.5 to 3.67 feet. This allows a quicker changeover and more manageable parts for manual labor.

When drilling starts, the kelly bar is almost completely above the wheel. As the drill cuts, the kelly bar and pipe will lower until the top of the kelly bar is level with the top of the wheel hub. Then the winch operator lifts the pipe until the slip plate can fit under the coupler and over the base (see Figure B.4). After unthreading the kelly bar from the drill pipe, the kelly bar is raised until it reaches the top of the cantilever

Figure B.4: Photo of the slip plate (yellow) holding a coupler and the drill string (extending vertically below it). The kelly bar is unscrewed from the drill string and lifted. A new pipe segment is added between the coupler and the kelly bar.

beam. A new 3 foot pipe section is then placed between the kelly bar and the pipe (see Figure B.4). The new section is threaded onto the pipe, and then onto the kelly bar. This is done under the wheel by one operator holding the pipe with a pipe wrench and the other operators tightening the kelly bar by turning the wheel (as the drill runs, the pipes will fully tighten). A wrench stop has also been welded to the base so that the operator does not have to supply the resistance to loosen or tighten the pipe. The pipe is lifted slightly, the slip plate is removed, the drill string is lowered, and drilling continues.

Pump

The final design of the drilling rig does not include a human-powered pump. The second major prototype made by the team included a prototype of a treadle pump system; however, due to time constraints, the design of the treadle pumps could not be more fully developed. In conjunction with the sponsor and team coach, it was determined that the necessary analysis and development of a pump would not be able to be satisfactorily completed in the allotted time. Although the treadle pump is still a feasible concept, designing and building a pump has been eliminated from the scope of the project.

Drilling Slurry and Pump Requirements

In order to operate an effective mud rotary drill, a drilling fluid must be utilized that can remove the cuttings from the borehole. This process occurs by pumping a viscous slurry down the hole through the center of the drill pipe. The slurry then returns through the annulus between the borehole wall and the pipe with the cuttings created by the drill bit. This process can remove any type of cuttings by adjusting the viscosity of the slurry. WaterDOT's drill design uses drilling fluid consisting of a mixture of bentonite and water, mixed at a ratio to provide the desired viscosity. Since the cuttings are typically denser than the drilling fluid, a combination of buoyant force and shear stress acts on the cuttings to propel them to the surface.

This results in pump requirements that can provide the necessary flow rate and fluid pressure. A flow rate of 50 to 100 gallons per minute is sufficient to create the necessary shear stresses on the cuttings and remove the cuttings at a quick enough rate. In order to provide adequate pressure, the pump needs to provide one foot of pressure head for every foot of depth of the borehole. This equates to a pressure of approximately 100 psi at a depth of 250 feet. Using these pump specifications, a pump power curve was created to determine the feasibility of operating a pump with human power. The resulting curve is shown in Figure B.5.

B.5 Exploratory Testing

The team's final design was only reached after considering and testing many ideas. Throughout the product development process, the team conducted tests to learn more about the difficulties of drilling and to determine if the selected concepts would meet the requirements. In all, the team has drilled holes 7 separate times, 2 as proof of concepts in Utah, 1 in clay in Utah, 1 in cobblestone in Utah, 1 in sand in Utah, one in rocky soil in Tanzania, and one in sandy soil in Tanzania.

Proof of Concept Testing

October 2010: The first idea that was proven through testing was the ability to turn the pipe by walking in circles around

Figure   B.5:
Pump   power
requirements.

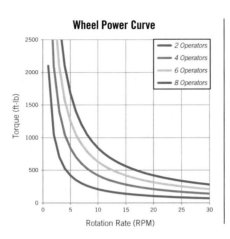

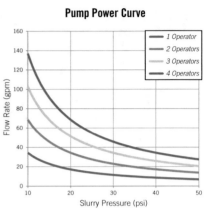

the pipe. The test was very simple. A drill bit was spot welded to a pipe, and using pipe wrenches to grip the pipe, the pipe was turned. During this test, 1 inch of depth was drilled in 10 minutes (see Figure B.6a). A similar test was performed with weights resting on top of the drill bit to provide downward cutting pressure (see Figure B.6b).

Before these simple tests, the team envisioned a system that would have the workers walk around the pipe twisting it as they walked in circles. However, while testing these early-stage prototypes the idea that it would be much easier to be stationary and pass the wrench around was developed. This idea was taken forward into the development of the first fully functional prototype.

November 2010: The first fully functional prototype was made of wood (see Figure B.6c). This was done to reduce cost and decrease manufacturing time. A six foot wooden wheel was used to harness human power to turn the pipe. This wheel had vertical handles and was pushed along by up to 6 workers that could stand around it in a circle. This design could be both operated with minimal effort and apply large amounts of torque to the drill pipe. This prototype was first tested in a small hole to ensure its feasibility. It met the team's expectations. The inertia of the wheel was able to keep the drill spinning in

between pushes. This made for a smooth operation. The diameter of the wheel was a good size to operate and it would easily enable operators to apply enough torque.

Drilling in Clay

December 2010: After the proof of concept tests, the fully functional wooden prototype was finished. The team faced some difficulty finding a location where property owners would allow holes to be drilled into the ground. The team used the property of one of the team members as the first place to test the prototype (see Figure B.6d). This location was selected because of ease of access to water and because the soil conditions were known (known to be clay). Parts of design that were specifically being tested were the pumping system, the wheel, and the amount of downward pressure needed to drill. Through 24 minutes of continuous drilling a hole 29 inches deep was drilled. From this test, the team learned that one treadle pump could not provide enough flow to lift all of the cuttings out of the borehole. This caused the drill to get stuck easily and increased the effort required by the operators to turn the wheel. When extra downward pressure was added the drill dug a little faster at first, but then the bit became stuck. It was determined that the ability to remove the cuttings needed to be improved by adding a second treadle pump before the next test.

Figure B.6: Field tests of drill system, in sequence of product development. a) Simple walk-around-in-a-circle proof of concept test in October 2010. b) Simple test with downward force and bit rotation. Test occurred in October 2010. c) Wood prototype for testing rotating bit by large wheel. Test occurred in November 2010. d) Wood prototype tested with treadle pump in clay. Test occurred in December 2010. e) Wood prototype tested in cobblestone soil with gasoline fueled mud pump. Test occurred in January 2011. f) Steel prototype tested in sandy soil. Test occurred in March 2011. g) Final prototype tested in rocky soil with local workforce. Test occurred in Tanzania in May 2011. h) Final prototype tested in sandy soil with local workers. Test occurred in Tanzania in May 2011. i) Use of refined production units by WHOlives.org.

Drilling in Cobblestone

January 2011: The next two tests were located at All American Gardens, a BYU property near the main campus in Provo, where the soil was known to contain rocks varying in diameter from 0.5 inches to 4 inches (see Figure B.6e). The team refers to this soil condition as cobblestone. These tests were performed on two separate days using the wooden prototype. In these tests a second pump was added and bentonite was used to thicken the drilling mud. This was done in hopes that the cuttings would be removed more effectively. However, during the second test both treadle pumps failed because they could not generate the pressure needed to move the thick slurry. To keep the test going, the team rented a mud pump and evaluated other aspects of the prototype.

The first 4 feet of digging was very similar to the test in clay, but then the cobblestones were encountered. The cobblestones made the drilling slow and arduous and it became difficult to measure progress. Since there was no way to lift the drill bit off the bottom of the borehole, the cobblestones were simply moved around instead of being cut through. Despite the slow progress, the prototype was able to drill through rock and pull up the cuttings with a mud pump. From the borehole a rock was pulled that had the profile of the drill bit deeply carved into it (see Figure B.7), and the settling pond contained shovels full of gravel. These were taken as proof that the drill had drilled through and removed rock (see Figure B.8). During these tests it became apparent that the design made it hard to access under and around the table to add and remove pipe. This resulted in modifications to the design.

The team decided that the final design needed to include a way to remove the wheel to provide greater access in and around the pipe interchange area. Also, the hoist should always be in place so that the pipe could be lifted and lowered while drilling. At this point it was also decided to remove the human-powered pump from the scope of the project. For the initial implementation of the drilling rig in

Figure B.7: Photo used as evidence that the drill was capable of cutting rocks.

Figure B.8: Photo used as evidence that the slurry was capable of lifting the cuttings out of the borehole.

Tanzania, a gas-powered pump will be used to pump the drilling slurry. Although this uses a consumable fuel, it will use drastically less fuel than a conventional rig.

Drilling in Sandy Soil

March 2011: The final test in the USA with the steel prototype was performed in Terra, UT, in sandy soil condition (see Figure B.6f). In all, 27 feet was drilled in 1.5 hours. The actual time the drill was spinning was 21 minutes. The average time for adding a new pipe was 2.5 minutes. Extrapolating from this data it is calculated that it would take 11 hours to drill 250 feet. This number may be optimistic, because it assumes that no problems will be encountered with increased depth that have not already been encountered. However, a professional driller who was present at the test stated that there is no reason to believe that it becomes harder to dig with increased depth. This makes the 11 hour estimate more feasible.

The ability to raise and lower the pipe while drilling was an important part of this success. When the drill string's (all pipes and cutting bit) full weight was resting in the hole, the drill would dig too fast and the wheel would become very hard to turn. The winch was used to control the rate of penetration. This made the drill easy to keep at a constant 30 RPM. Being able to keep a constant rhythm while spinning the wheel greatly increases the human power sustainability.

Before this test, the process of adding new pipe had only been tested once at the All American Gardens. The procedure was very difficult, dangerous, and took the entire team to perform. One of the main purposes of this test was to evaluate our pipe changing procedure. In the current version of the design, the team made the pipes smaller, for easy handling, and cleared out space to work underneath the wheel. During the testing it was very easy to change the pipe with only two people. Overall the results were very pleasing.

Drilling in Rocky Soil in Tanzania

May 2011: The first test in Tanzania used a refined version of the steel prototype described above. The test was performed on the outskirts of Arusha Tanzania (see Figure B.6g). The goal of this dig was to evaluate the challenges of training the local workers how to use the machine. It was also to see how the drill would do digging through rock formations. The team was fortunate to encounter a professional well driller, who showed us where to dig to encounter rocks. The first few meters of digging were simple and completed without difficulty. The team encountered rock, just as predicted, and the drilling slowed significantly. At its slowest the drill cut 0.75 inches of rock in one hour of continuous

drilling. The team broke through some rocks and encountered others. On the most difficult day of cutting through rock the team cut 53 inches of rock. A 27 foot hole was created as part of this test.

In the end, the local workers had little to no trouble understanding the drill and how to use it. Also, the drill continued to dig through rock the entire time, albeit slowly in some cases. The team showed that the drill can indeed dig through rock at a rate that is consistent with the client's goals.

Drilling in Sandy Soil in Tanzania

May 2011: The final test performed by the team was to dig as deep as possible in the time remaining in Tanzania. The goal was to test the functionality of the pump at depths beyond 100 feet, and to observe the difficulties that exist the deeper the bit travels.

The team found a sandy soil formation a few hours outside of Arusha Tanzania, in the small town of Magugu (see Figure B.6h). The dig through sand was simple and straight forward. After just a few hours, the local workers were handling the entire operation. The pump continued to work beyond the depth the team had predicted it would fail. The team had planned to put two pumps in series, but this was not necessary.

From the test, the team observed that the structure worked well and that with small improvements it would have no trouble meeting the client's requirements. The part that began to yield in the design was the shaft that held the pulley at the top of the beam. This particular part of the product needs to be refined. Also, by standard drilling procedures, the drill string should be pulled up each night to avoid seizing the bit. An unanticipated issue related to this is that the deeper a team digs each day, the more time needs to be reserved to remove the drill string, and the more time is needed the next morning before actual digging can begin since the string needs to be lowered into the hole.

In the end, a 140 foot hole was created as part of this test.

B.6   Team Conclusions and Recommendations

Throughout the course of the development process many improvements have been made to the design to ultimately create a design that can efficiently drill a borehole. Through testing, it has been determined that the current design is capable of drilling in several soil types including clay, sand, loose cobblestones, and rock. Although at times the progress may be slow drilling through rock, the drill remains capable of cutting. The drill is also affordable, easily transportable, and robust, meeting the main requirements.

Although team WaterDOT has successfully met the functional specifications, there is still room for the design to be improved. There are three main areas that can be considered for improvement: manufacturability, safety, and cost. As a whole the manufacturing of the device is accomplished with simple operations; however, there are a few components that are manufactured using mills. Ways to eliminate the need for these more complex operations can be sought. The device also contains many exposed moving parts, which can be better shielded to prevent the possible pinching of operator's body parts. Finally, ways to reduce the overall cost of the device should be explored.

In addition to these three main areas of overall improvement, the team recommends that tool joints be used at every pipe connection to improve change-over and prevent over-tightening of joints. Also, a second slip plate can be added to introduce redundancy to better prevent the pipe from falling down the borehole during the removal of pipe. A sealed thrust bearing can be used between the wheel hub and the wheel support to protect against corrosion and to improve the performance of the wheel. Finally, the wheel can be improved by employing a unidirectional mechanism that can prevent the wheel from being spun in the wrong direction and employing a method of stopping the wheel while it is turning.

Team WaterDOT has successfully met the requirements of designing and building a human-powered water well drill that is capable of reaching underground potable water. This success is measured by meeting the functional specifications. The final prototype was manufactured for $1,600 USD and is easily disassembled into pieces that can fit in a truck bed or a small trailer. Furthermore, the rig is able to successfully drill in many soil conditions. In all, the team has developed a product that can be successfully implemented in Tanzania and bring clean water to villages, improving the quality of life for potentially thousands of people.

Product Development Terminology

Terms used in this book are defined below. While we have tried to keep definitions consistent with accepted usage, in some cases words have several different meanings.

Our usage throughout the book is consistent with these definitions.

Acceptable limits (n): Limits on the value of a performance measure that define the boundaries of a desirable product. If a product has performance above the upper acceptable limit or below the lower acceptable limit in a given measure, the product will not be desirable, regardless of how well it performs in other measures.

Approval (n): A statement given by the project's approvers that the desirability and transferability of the design are sufficient to move on to the next stage of development. Approval will generally be given the design has been transferably demonstrated to meet market requirements through the evaluation of artifact checks, performance testing results, and validation testing results.

Approver (n): A person or organization charged with granting approval at the end of a development stage.

Approvers often include the client and organizations in the development company outside the development team. Very rarely will the market act as an approver; the market input is obtained during validation testing.

Architecture (n): The arrangement of and interface between major parts and subsystems of a concept or system. As used in this book, the product architecture consists of a concept, a definition of its major subsystems, and a definition of the interfaces between subsystems.

Artifact (n): An object made by the product development team as a transferable result of their work. Also referred to as a *product development artifact.* In general, there are three classes of product development artifacts: requirement artifacts, test artifacts, and design artifacts.

Bill of materials (n): A design artifact, in the form of a table, listing each component comprising a design. The bill of materials generally includes information beyond the name of the part; for example, it can include the material of each part, the cost of each part, part number, approved vendors, and so on. It is often organized in a way that makes it clear which

© Springer Nature Switzerland AG 2020
C. A. Mattson, C. D. Sorensen, *Product Development,*
https://doi.org/10.1007/978-3-030-14899-7

components comprise each subassembly.

Checks (n) as in *artifact checks:* The evaluation of the extent to which a product development artifact representing requirements, tests, or design is complete, unambiguous, transferable; accurately conveys the nature, scope, or meaning intended by the product development team; and complies with applicable artifact standards.

Client (n): The entity paying for or otherwise driving the product development effort. In most cases it is someone other than the *end user.*

Component (n): A low-level, often lowest level, part in a subsystem or subassembly.

Concept (n): A means for achieving a design outcome or meeting a design requirement, where enough detail is provided about the spatial and structural relationships of the principal subsystems that basic cost, size, weight, and feasibility estimates can be made.

Coordination interval (n): The smallest period of time between team or subteam coordination actions.

Customer (n): An individual who is likely to purchase or has purchased the product. A customer is a member of the market, and may be part of the market representatives. The customer is often an *end user.*

Decomposition (v): The act of breaking something down into simpler constituents.

Design (n): The definition of the product that will eventually be transferred to the production system and used to determine what product is made. The design also includes any testing procedures and performance measures necessary to verify the quality of the produced product. The design evolves throughout product development along with *requirements* and *tests.*

Design (v): The act of advancing a *design (n)* through the stages of development. Note that to avoid confusion, we do not use the word design as a verb in this book.

Design activities (n): Actions the development team or development team members take to advance the product design through the stages of development.

Designer (n): One who engages in product development activities. Also called product developer.

Desirable (adj): Wanted or wished for as being attractive, useful, or necessary. Used here to mean that the quantitative and qualitative product performance are sufficient to entice customers to purchase the product.

Idea (n): An indefinite or unformed conception.

Ideal values (n): The values of the performance measures for a product that the market would like to have, neglecting any required trade-offs. Ideal values generally comprise an ideal value and one or both of an upper acceptable limit and a lower acceptable limit.

Market (n): The group of customers who are interested in the purchase of a product.

Market opportunity (n): The set of requirements that define what the market would like to have in a product. The market opportunity also comprises performance measures for the requirements, requirement–measure relationships, and ideal values.

Market representative (n): An individual or group of people chosen to represent the market for purposes of defining and evaluating product desirability. In most cases, the best designs result when the market representative consists of individuals outside of the development team.

Market response (n): A statement given by the market or the market representatives as to the desirability of a product relative to the market requirements.

Market requirements (n): Objective and subjective product characteristics that determine the desirability of the product to the market.

Measured value (n): The value of a performance measure that has been demonstrated by testing a prototype.

Outcome (n): The result of a design activity. In most cases, outcomes in activity maps describe the state of the design (or of particular artifacts) that is expected at the end of the activity leading to the outcome.

Performance measure (n): A characteristic of a product that will be measured by the development team during performance testing. This evaluation is used as an internal measurement that can give guidance about the actual market preferences.

Performance test artifacts (n): Artifacts used for performance testing. These artifacts typically include prototypes and models based on the current design.

Performance testing (n): The evaluation of the performance of a design, prototype, or product. Measured or predicted values of performance measures are obtained by performance testing. Evaluation of the design quality is partially based on how well the tested performance meets the design targets.

Post-release refinement (n): An engineering activity that refines the product after it has been introduced to the market, including correction of post-release defects, incompleteness, documentation, producibility, reliability, or any other issues that are directly tied to product desirability and transferability.

Predicted values (n): Values of performance measures that a product is expected to have based on testing of predictive models.

Producibility (n): The measure of the relative ease of manufacturing a product that meets market requirements. A highly producible product should be easily and economically fabricated, assembled, inspected, and tested with high quality.

Product (n): A tangible item manufactured by a production system and purchased by someone to fulfill some need.

Product development (n): The act of advancing a design through the stages of development.

Product development process (n): The plan for advancing a design through the stages of development.

Project (n): An organized effort to execute the product development process, or to advance the design through one or more stages of development.

Project management (n): The act of guiding a development project to a successful conclusion. Project management usually consists of tracking the progress of the design and comparing it with the project plan, then adjusting activities and resources to ensure timely completion of the desired outcomes.

Project planning (n): The act of choosing which design activities to execute, when to execute them, and with what resources in order to successfully complete a project.

Prototype (n): A physical approximation of the product, in whole or in part, based on the design at the time the prototype is created.

Real values (n): Values for performance measures that are chosen by the team to best meet market requirements. The real values include the effects of trade-offs, recognizing that it is generally impossible to meet every market requirement. Real values include a target value, predicted values, and measured values.

Requirements (n): Product development artifacts that capture the definition of desired and measured performance for a product. The requirements comprise market requirements, performance measures, requirement–measure correlations, ideal values, and target values. They also comprise the predicted and measured values and market response.

Requirements matrix (n): A matrix used to display the requirements in an integrated way.

Stage of Development (n): A portion of the development lifecycle where particular effort is placed on creating a specific set of product development artifacts. A stage is completed when approval is given after checking the artifacts to determine transferability and testing and validating the performance to determine desirability.

Stakeholders (n): A person or set of persons with an interest or concern in the outcome of the development.

Subsystem (n): A self-contained system forming a subset of a larger system.

System (n): A set of connected things or parts forming a complex whole. For the purposes of this book the system is the whole design or the whole product.

Target values (n): The performance measure values hoped-for after considering the trade-offs between the various market requirements. Because the choice of target values takes into account the trade-offs across performance measures, they can only be chosen after a concept has been selected. This is because the nature of the concept determines the nature of the trade-offs.

Tests (n): Methods used to predict or measure the performance of a product or its competitors. Tests will generally

be applied to prototypes for testing measured performance and to models for testing predicted performance.

Transferable (adj): Able to be used for its intended purpose when delivered to a recipient who has no specific knowledge except that which is transferred.

Validation (n): The assurance that an artifact meets the expectations of the market. It often involves evaluation by a market representative (which is always someone external to the product development team).

Bibliography

Altshuller G, Shulyak L (1996) And suddenly the inventor appeared: TRIZ, the theory of inventive problem solving. Technical Innovation Center, Worcester

Avallone E, Baumeister T, Sadegh A (2007) Marks' standard handbook for mechanical engineers, 11th edn. McGraw-Hill, New York

Boothroyd G, Dewhurst P, Knight W (2010) Product design for manufacture and assembly. Manufacturing engineering and materials processing, 3rd edn. Taylor & Francis, Boca Raton

Brooks F (2010) The design of design: essays from a computer scientist. Pearson Education, London

Brown T (2009) Change by design: how design thinking transforms organizations and inspires innovation. Harper Business, New York

Budynas R, Nisbett J, Shigley J (2011) Shigley's mechanical engineering design, SI version. McGraw-Hill series in mechanical engineering. McGraw-Hill Education, New York

Chase K, Parkinson A (1991) A survey of research in the application of tolerance analysis to the design of mechanical assemblies. Res Eng Des 3(1):23–37

Cushman W, Rosenberg D (1991) Human factors in product design. Advances in human factors/ergonomics. Elsevier, Amsterdam

Dieter G (2000) Engineering design: a materials and processing approach. McGraw-Hill series in mechanical engineering. McGraw-Hill, New York

Dym C, Little P (2008) Engineering design: a project based introduction. Wiley, Hoboken

Griffin A, Hauser JR (1993) The voice of the customer. Market Sci 12(1):1–27

IDEO (2003) IDEO method cards: 51 ways to inspire design. IDEO, Palo Alto

Jiao JR, Simpson TW, Siddique Z (2007) Product family design and platform-based product development: a state-of-the-art review. J Intell Manuf 18(1):5–29

Kelley T, Littman J (2001) The art of innovation. A currency book. Doubleday, New York

Miklosovic DS, Murray MM, Howle LE, Fish FE (2004) Leading-edge tubercles delay stall on humpback whale (Megaptera novaeangliae) flippers. Phys Fluids 16(5):L39–L42

Norton R (2006) Machine design: an integrated approach. Pearson Prentice Hall, Upper Saddle River

Osborn A (1963) Applied imagination; principles and procedures of creative problem-solving. Scribner, New York

Pahl G, Beitz W, Feldhusen J, Grote K, Wallace K, Blessing L (2007) Engineering design: a systematic approach, 3rd edn. Springer, London. https://doi.org/10.1007/978-1-84628-319-2

Pugh S (1991) Total design: integrated methods for successful product engineering. Pearson Education. Addison-Wesley, Boston

© Springer Nature Switzerland AG 2020
C. A. Mattson, C. D. Sorensen, *Product Development*,
https://doi.org/10.1007/978-3-030-14899-7

Sanders MS, McCormick E (1993) Human factors in engineering and design. McGraw-Hill International Editions, 7th edn. McGraw-Hill Higher Education, New York

Shah JJ, Smith SM, Vargas-Hernandez N (2003) Metrics for measuring ideation effectiveness. Des Stud 24(2):111–134

Tilley A, Henry Dreyfuss Associates (2002) The measure of man and woman: human factors in design, vol 1. Wiley, New York

Ulrich K, Eppinger S (2012) Product design and development. McGraw-Hill, New York

Young W, Budynas R, Roark R, Sadegh A (2012) Roark's formulas for stress and strain, 8th edn. McGraw-Hill Education, New York

Index

Printed in the United States
By Bookmasters